The True Path

The True Path

Western Science and the Quest for Yoga

Roy J. Mathew, M.D.

PERSEUS PUBLISHING
Cambridge, Massachusetts

Cataloging-in-Publication Data is available from the Library of Congress.
ISBN 0–7382–0458–7

Perseus Publishing is a member of the Perseus Books Group.
Find us on the World Wide Web at http://www.perseuspublishing.com.
Perseus Publishing books are available at special discounts for bulk purchases in the U.S. by corporations, institutions, and other organizations. For more information, please contact the Special Markets Department at The Perseus Books Group, 11 Cambridge Center, Cambridge, MA 02142; or call 617-252-5298.

Text design by Cynthia Young
Set in 11-point Adobe Garamond by Perseus Publishing Services

First printing, May 2001

1 2 3 4 5 6 7 8 9 10—03 02 01

This book is dedicated to alcoholics and addicts and people who care for and about them.

Contents

Acknowledgments

This book, different from anything I have thus far written, turned out to be much more laborious than I ever anticipated. Several people helped, supported, and encouraged me in my effort. I discussed my thoughts and ideas with a number of friends, scientists, theologians, and philosophers and consulted a number of books, both in philosophy and science. Among the many who provided invaluable help to me, a few names stand out.

My sourcebook for Indian philosophy was the two-volume work on the subject by Dr. S. Radhakrishnan, the noted Indian philosopher and one of the past presidents of independent India. I found myself going back to his writings over and over again to clear up issues I could not fully comprehend. I have borrowed freely not only from Radhakrishnan's philosophical ideas but also from his literary style. My wife, Laura, and son, David, patiently endured the long hours of my absence when I was working on the book. Throughout my life my older brothers, V. George Mathew and John B. Mathew, and my close friend T. L. James have stood by me wherever I went and through whatever mistakes I made. They have been of invaluable assistance to me in this enterprise as well. My oldest brother, V. G. (as we call him), in his characteristic unassuming style, offered a number of criticisms that I resisted at first as I always seem to do. In the long run, every single one of his suggestions turned out to be meaningful, insightful, and incredibly useful. I have had discussions with Reverend William Willimon, dean of Duke Chapel, on many relevant topics, especially concerning the Christian faith. He was always willing to take time out of his busy schedule to listen to my ramblings on Hinduism and its relationship with Christianity. He cleared up a number of doubts and helped me formulate my thoughts.

I first became interested in this topic through my interactions with alcoholics and addicts. In the early days, when I received minimal encouragement from the scientific community, my patients, friends, and colleagues were always there to keep my interest up. I owe them, especially Jeff Georgi of the Duke Addictions Program, a large debt of gratitude. Kathryn Elliott Riffle, another friend, was also a constant source of support and inspiration. William H. Wilson, my partner in research, helped me by bringing up the "other" point of view.

I would like to acknowledge the following organizations and individuals for their generous efforts and contributions: Lippincott, Williams and Wilkins for the use of brain images from their LifeART series, the journal *Electroencephalography and Clinical Neurophysiology* for the use of the diagrams in Chapter 9, Vicki Russell for her proofreading skills, Brenda White and Judy Ridley for their typing talents, and Robin Davis for her coordinating and secretarial efforts. I am also very grateful to Professor Berndt Mueller, Department of Physics, Duke University for helping me with the parts that deal with physics. Ms. Val Lauder, from the School of Journalism and Mass Communication at the University of North Carolina at Chapel Hill, helped me organize my thoughts and ideas and express them in a manner interesting to the reader and acceptable to the publisher.

Preface

Since I came to the West, one of the questions I am asked most often is why I have a name like Roy Mathew. Many suspect that my family converted into Christianity under the influence of the British when they were in India, but this is not correct. We have been Christians for more than 1,500 years, long before the British arrived in India. Syrian Christians, as we are called, believe the apostle St. Thomas came to India after the crucifixion to bring the news of the long-awaited Messiah to the Jewish community in southern India. Some Christian families accompanied Thomas, and several more came a few hundred years later. Thomas was also successful in converting some of the locals into the new faith. When the Europeans first arrived in India, they were surprised to find crosses and churches.

The Europeans did successfully attempt to convert some of the original Christians to Catholicism and the other European Christian factions. However, many families remained faithful to their ancient tradition. Our family was among them. Under the English influence, our names, which were originally closer to Jewish names, were Anglicized. My baptismal name, Yakob, became Jacob in official records, and after I came to the United States, it shrank into a "J" as my middle initial.

India is a mosaic of different faiths and traditions. Hinduism, the most popular religion—if it can be called one—is remarkable for its diversity. It has never been a single, homogeneous entity. It comprises a wide-flung, motley collection of overlapping beliefs, principles, and practices that are divided further by the ever-present caste system. Hindus, until very recently, were not threatened by other faiths; in fact, they enthusiastically welcomed and wholeheartedly embraced fresh thoughts and ideas. Buddhism, Jainism, and Sikhism originated in India, whereas Christianity, Islam, Zoroastrianism, and, more recently, Bāhāism came from abroad. Thus, for a Christian, growing up in India invariably meant exposure to other religions and faiths, especially Hinduism. Hinduism is much more than a religion; it is intimately related to the entire Indian culture and civilization. Most arts and sciences have close ties with Hinduism. Regardless of one's religious orientation, all Indians—including Christians—accept, at least to some extent, Hindu social mores: the caste hierarchy, arranged marriages, and astrology.

Most of my early life, until I finished my undergraduate medical training, was spent in India. I too accepted the unwritten rules of social life in India. I went to England for postgraduate medical training and from there, in search of better opportunities for a career in academic medicine, to the United States. My life in the West brought up a number of new issues.

In India, philosophy and religion are intimately intertwined; there is no philosophy without a religion and no religion without a philosophy. In the West, on the other hand, philosophy and religion have a much looser relationship. Many prominent Western philosophers are not very religious; some are not religious at all. Most well-known scientists are either atheists or agnostics, and the others wall off their religious beliefs from science.

When I came to the United States, the training I received in England in research methodology stood me in good stead and helped me launch a career in neuroscience research. My early research had no relation to philosophy at all. Like my colleagues, I too carried out studies based on rigorous principles of science and published papers in scientific journals. For many years, I was comfortably settled in my career as a neuroscience researcher. As I became more and more engrossed in my work, I pushed my philosophical interest to the back of my mind.

It was my work with drug addiction that turned me around. I came across large numbers of substance abusers in recovery who traced their recovery back to a "spiritual experience." While there was no firm scientific support for their claim, I simply could not overlook the testimonies given by numerous individuals. This rekindled my interest in philosophy and brought back to life thoughts, ideas, and images from my distant past in India. While I believe in research methodology and admire the many benefits scientific research has brought to mankind, philosophy and religion have also contributed substantially to human welfare. Both are equally important, but a deep gulf has separated them.

This book is my modest attempt to bridge the chasm between science and philosophy. The book by no means provides all answers, and it is far from being complete and comprehensive. There are any number of gaps, voids, and omissions. However, it does represent my firm belief that ultimately everything is interrelated and reducible to one underlying principle.

The book attempts to bridge philosophy and science, and that, indeed, is a tall order. In my wildest dreams I could not hope to cover anything but an infinitesimal fraction of these two large bodies of human thought and effort. Many schools of philosophy and important religions are not adequately discussed here, and many are not even mentioned. This in no way takes away from their usefulness, meaning, or significance; it simply reflects limitations of my knowledge and familiarity. Although very little in this book is totally incompatible with science, strictly speaking, this is not a textbook on science. Within these pages, speculations overwhelm established facts. Once again, I could not discuss in depth the relevant, difficult, and controversial topics in neuroscience because of space limitations and complexity. The book is primarily meant for general readers, but I hope both philosophers and scientists will find it of interest.

Although there are a number of books on the brain and spirituality, this book focuses primarily on Indian philosophy. In the West, there is an unfortunate tendency, both in science and philosophy, to recognize and acknowledge only Western authors, of both past and present. It would appear that all human thought took root in Greece. For example, Hippocrates is often touted as the father of scientific medicine, and his contemporaries in India, Caraka, and Śuśruta, who wrote more voluminous and in-depth treatises on the subject, are omitted. Similarly, René Descartes is credited with championing the mind-body dichotomy, while the Indian thinker, Kapila and his system of philosophy, Sāmkhya, which preceded Descartes by at least a thousand years, is sadly left out. Thus, the student is deprived of information of considerable sum and substance and the original authors, ignored. I hope that my book will bridge this gap, at least to some extent.

Introduction

Conversion Experience

"The Lord came to me last night," the young woman said with excitement. I looked up from the medical record in which I was entering the day's progress notes. Admitted the previous week to the Duke University Medical Center for treatment of addiction to cocaine, Mary Smith had been brought to the emergency room by her aunt after she tried to commit suicide with a large dose of aspirin. After resuscitation and stabilization, she was transferred for continued treatment to the substance abuse ward where I was medical director.

Several years before her suicide attempt, a friend had introduced her to crack cocaine. Cocaine, which used to be expensive and relatively unavailable, had become cheaper and readily available. Mary had heard a great deal from her friends and acquaintances about the intense euphoria the drug produced, and she was curious to try it—once—for the experience. Too scared and embarrassed to find the drug herself, Mary accepted a friend's offer to obtain the cocaine for her. Her friend even offered to pay for it and show Mary how to use it. One evening, Mary smoked a small quantity of crack cocaine with her friend in her house.

A few weeks later she wanted to try it again.

What started as recreational use, over time, grew into a firmly established addiction that she could not shake. Obtaining the drug was not a problem in her neighborhood. She started spending most of her salary as a secretary on cocaine, which resulted in her family of two young children going without their basic needs being met. Soon, Mary was spending $100 to $200 per day on crack cocaine, an expense her modest salary could not support. The addiction progressed relentlessly, forcing her to borrow money from her friends and relatives, which she was in no position to repay. When they started avoiding her, she resorted to shoplifting. Shortly after, she was arrested and convicted, after which she lost her job.

By then she knew that she was on the road to total destruction, and she desperately wanted to stop. She had become her own worst enemy, fighting the addict within herself, only to lose the fight every single time. Soon, she ceased to struggle. Obtaining and using cocaine became her single-minded goal, eclipsing every single dictum and scruple in her life. There was nothing, not even the agony of her children, that could stop her, and there was nothing she would not do to get cocaine. She sold all her belongings, including her car, television, and refrigerator, to buy cocaine. Once she'd exhausted everything she owned—including even her children's clothes—she started selling herself. She prostituted in local crack houses in exchange for crack cocaine, despite the moral degradation and the risk of venereal diseases, including HIV. Then she contracted gonorrhea. At this point, Mary had reached the end of her rope. She had become a social pariah: unemployed, penniless, and without an iota of self-respect. She stopped going to church. In fact, she actively avoided thoughts of God and religion.

Late one night, on her way back from the crack house, two strangers abducted her. They took her to an abandoned house where, at gunpoint, they brutally raped her multiple times. After they were finished with her, they threw her from a moving car. It was raining hard. Bleeding, bruised, and half naked, she walked home in the dark, drenched to the skin. Because she had not paid her electricity bills, there was no power in the house. As she sat in virtual darkness and gazed at her two neglected children fast asleep on a filthy bed, she hated herself and what she had become. To her, the only way out was suicide.

Despite my best efforts at consolation and encouragement, I could not bring her out of the deep, dark morass of depression that engulfed her. In her opinion, she had crossed the point of no return and there was no hope for her. She had transgressed every doctrine and dictum, of man and God. Even if others were willing to forgive her, she herself could never condone what she had done to herself and to her children. Counseling was made difficult by her extreme reluctance, perhaps inability, to face herself. I was seriously concerned about a second, more determined suicide attempt.

Yet this morning she appeared before me full of hope and excitement. Her eyes, red, swollen, gloomy, and tear-drenched the previous day shone like stars with happiness. I was pleased but perplexed. Not strongly religious myself, I searched for a psychiatric explanation for this dramatic transformation. I did not know what to make of her claim that the Lord had come to her.

"I'm not sure what you mean," I replied.

"The Lord was with me last night, Doctor."

"Now, come on Mary. Are you trying to tell me that the Lord was with you in this room last night? What did he look like?"

"No, I didn't see the Lord like I see you, but I know the Lord was with me. I felt it. I felt it in my bones. I wish I could describe it to you better, but I just can't."

"I know it's going to be all right," she said, "I know I won't be using drugs again."

Whether she actually saw the Lord or not, there was a very visible change in Mary. Clearly, some event had taken place the previous night that elevated her from the murky depths of melancholia to buoyant excitement and optimism. This was strange indeed. I questioned her further.

"Tell me exactly what happened."

Mary responded:

Last night I couldn't sleep. I never felt more miserable in my whole life. When you're out there, busy getting high on cocaine and figuring out how to get more cocaine, you don't think about yourself. In a crack house you don't think about right and wrong.

You have only one thing on your mind—getting high. In the hospital, surrounded by good people, with empty time on your hands, you come face-to-face with yourself, face-to-face with your conscience. I saw what I'd done to myself and my children. I was a miserable mother and a miserable woman, worse than scum. God wouldn't give me even death. I wasn't even fit to die. I fell down on my knees and begged for mercy. I cried like I had never cried before. I don't know how long I remained on my knees. Suddenly, something happened to me. Something snapped. In an instant, the pain was gone and I felt better. The entire room seemed to brighten up and I knew that I wasn't alone. I knew everything was going to be all right.

I was very happy for Mary. Her conviction, whatever it was based on, was definitely to her benefit. How long it was going to protect her from cocaine, I did not know. Recovering from a firmly established cocaine addiction such as Mary's is by no means an easy task.

Approximately six months after Mary shared her experience with me, I was standing by the glass window of a car wash marveling at the amazing job the machines were performing in cleaning my extraordinarily dirty minivan. A system of chains and pulleys dragged my minivan, encrusted with dirt and muck, along metal tracks and subjected it to the torments of jets, which sprayed it with soap and hot water; rotating brushes, which scrubbed it clean; fountain heads, which sprayed wax on it; and hot air blowers, which dried it. The end product that emerged clean and sparkling was a far cry indeed from the muddy mess that went in.

While engrossed in watching this event, a woman I could not quite place startled me by coming up to me and encircling me in an embrace.

"You don't know me, do you?" she asked.

Considerably embarrassed by her public display of affection, I admitted my inability to recognize her, although she looked very familiar.

With eyes full of pride and happiness, the woman exclaimed: "I'm the person who saw the Lord six months ago on your ward. I told you I wouldn't use drugs again and I haven't."

In the conversation that followed, I discovered that in the months following her release from the hospital, Mary Smith had been working at that car wash and had indeed remained drug free. The manager of the car wash, whom I knew well, confirmed her report. Mary's mysterious experience with the Lord seemed to be the only treatment she needed to kick her cocaine addiction.

Spirituality and Science

As a psychiatrist and as a scientist, I could not offer any logical explanation. None of the scientific theories propounded to explain the phenomenon of addiction could explain her recovery. A mysterious event of a few minutes, in the dead of night, had released her completely from the shackles of addiction. In addition, it had erased from her mind all stains and scars of addiction and had brought serenity and happiness back into her life.

My scientific mind had considerable difficulty in accepting this. Perhaps it was a placebo effect, a chance occurrence. Although foreign to the scientist, such happenings are well known to individuals with addictions and the clinicians who treat them. Alcoholics Anonymous (AA) and Narcotics Anonymous (NA) are based on the notion that a spiritual experience will deliver sufferers from the maws of addiction. In fact, Bill Wilson, the founder of AA, had an experience similar to Mary's.

Apparently, hundreds of thousands of alcoholics and addicts have found their recovery through spiritual experiences. Although AA is the single most popular approach to the treatment of addiction, scientists have, by and large, brushed aside AA and spirituality. Yet Mary Smith and her recovery were real.

Although I had not previously encountered such experiences firsthand, I had certainly read about them. Bill Wilson, perplexed by his experience, consulted psychologist William James's renowned book *The Varieties of Religious Experiences.* According to James, who subsumed such experiences under the title "Conversion Experiences," they blossom out of the murky depths of despair and depression. The individual is usually transformed for life.

Such experiences are by no mean limited to addicts. Siddhārtha, known as Sākhya Muni in ancient India and as the Buddha elsewhere, had such an experience. Siddhārtha, the crown prince of the Sākhya kingdom in the foothills of the Himalayas, was born in 567 B.C. Young Siddhārtha, gifted with an extraordinary degree of sensitivity, became consumed with the misery of human existence: of disease, aging, and death. Unable to find peace until he had found an answer to human suffering, the young prince left his wife and infant and wandered out into the dead of night. He renounced the comforts of the palace, donned the saffron robe of an Indian ascetic, and traversed the country on foot seeking a solution. He approached several spiritual masters and, under their tutelage, put his mind and body through the rigors of asceticism. After six years of severe self-abnegation, his health deteriorated and he came very close to death. Yet the answer he sought continued to elude him; he remained unfulfilled and empty. Frustrated but unwilling to give up and give in, the young Buddha adopted a different approach, a middle path that avoided extremes. One night, while seated under a large pipal tree in a dense forest in Bodhgaya, the long-awaited answer came to him in a flash of insight.

The apostle Paul on his way to Damascus was transformed by a similar experience. Paul was actively engaged in persecuting followers of the newly heterodox offshoot, Christianity. Possibly deep down, he was tormented by guilt feelings. On his way to Damascus, a blinding "light from heaven" shone on him; it changed Paul forever. Instead of being Christianity's relentless persecutor, he became one of the most prominent proponents of the faith.

The prophet Muhammad, while praying and meditating in the Cave of Hira on Jabal Nur overlooking Makkah, had an extraordinary, paranormal experience. He was contacted by the angel Gabriel. After that encounter, Muhammad, who was illiterate, produced the Qur'an (Koran), the jewel of Arabic poetry, a work of great philosophical and spiritual depth.

If I were to dismiss Mary Smith's experience as illusory, then I would also have to dismiss the writings of William James and Bill Wilson, not to mention works by the founders of most religions of the world. If I were to dismiss Mary Smith's experience as invalid, then I would also have to dismiss the Buddha, Paul, and Muhammad.

Experiences like Mary's are admittedly outside the bounds of conventional science. Does that mean, however, that they are unimportant? Clinically, they unquestionably are of pivotal significance. Such experiences form the basis for AA, the backbone of substance-abuse treatment, with 15 to 20 percent of the U.S. population afflicted by addictive disorders. Religious people are known to cope better with stress, and they recover from such mental illnesses as depression more rapidly. Faith has been found to be beneficial even with such physical conditions as heart attacks, strokes, immune disorders, and open-heart surgery. Yet scientists have been squeamish about this topic.

If one were to conclude that a living brain is, indeed, a necessary prerequisite for a religious experience, it should not be hard to accept the notion that something should happen *in* the brain in association with the religious experience as well. The important question is: Which part of the brain might be involved in the generation of such an experience and what change might have taken place in that brain region?

The brain is the seat of all mental operations, and of life itself. The full spectrum of human experiences, fact and fantasy, sense and nonsense, physiology and philosophy, are all contained within the gray folds of brain tissue. Past brain research was, by and large, restricted to the study of topics relevant to neurology, psychiatry, and conventional behavioral sciences. However, the seemingly limitless landscapes of the mind clearly extended beyond the bounds of objectivity and conventional science. With the techniques now available for brain research, it should be possible to go beyond the pale of conventional neurosciences into the large expanses of misty, ill-defined regions hitherto unexplored and uncharted.

My main area of brain research involves functional neuroimaging. In the normal brain, cerebral blood flow and metabolism are tightly coupled to brain function. Thus, activation and deactivation of various brain regions can be identified and evaluated using cerebral blood flow and metabolism. The advent of noninvasive techniques for the measurement of cerebral blood flow and metabolism has made the study of regional brain activity in normal individuals possible. Over the years, my research team and I have conducted many studies on the effects of different drugs on cerebral blood flow and metabolism. With my newfound interest in spirituality, I reexamined our research findings. Since time immemorial, in various cultures, drugs have been used for spiritual enhancement, and a number of drugs we had experimented with—especially marijuana—were believed by many to enhance spiritual feelings. The data we had already collected provided new insights into relevant brain mechanisms.

Most of the available literature on spirituality tends to be vague and unfocused. The writings of most authors, even several influential ones, did not help me much in my effort to apply scientific techniques to study spiritual issues. To me, sound reasoning and logic, though not data based, were essential. It should be noted that Darwin's celebrated theory of evolution was based on keen observation and rational explanation and not on statistical analyses of numerical data.

Philosophical viewpoints presented by several Indian philosophers, especially Śāṁkara (A.D. 788–828), had a firm footing in reason. His descriptions of the human mind and its altered states were extremely helpful to me in understanding aspects of the spiritual experience in neurophysiological terms. His writings on metaphysical issues paralleled contemporary findings in quantum physics and astrophysics to a very considerable extent. Śāṁkara, who was born in the same part of India that I was, had a special appeal for me. Very soon I was enamored with him and his Advaita Vedānta (philosophy of nonduality). That was my stepping-stone to Indian philosophy.

Indian philosophy is vast and very diverse, encompassing the most complex and the simplest, the most profound and the most superficial, the broadest and the narrowest concepts. I started out with a cluster of ideas that made sense and were internally coherent. New information relevant to, and consistent with, the core ideas was added when it became available. The model I came up with was supported, at least in part, by what I knew in neurophysiology and neuroanatomy.

In India—and elsewhere—God means different things to different people. While some relate to an anthropomorphic image, the more philosophically minded conceptualize the Divine in abstract terms. According to many religions, God created the universe. However, many Indians believe that the creator transformed into the creations and that the creator and the creations are essentially one and the same. In sum, creations are distortions of the absolute caused by time and space.

Albert Einstein showed that time and space were not absolute and that they undergo changes. These alterations cause the manifested world, as we know it, to undergo drastic changes in form and shape. Thus, reality as we know it rests on the unstable pillars of time and space.

Indians see life as essentially a burdensome ordeal where losses dominate. Sickness, aging, and death bedevil life. That which is born must die; escape from death is not possible without escape from birth. Salvation is escape from the pain of the birth-death cycle.

Although the idea that time and space are unstable is supported by physics, the realm past these dimensions is beyond science. Human perception is dependent upon time and space, and all objects we know of are, therefore, dependent upon these two entities. An existence independent of time and space is beyond perception. The only way to know it is to experience it. Experience of the absolute means escape from time and space. Since this experience must take place in—or at least involve—the mind, it must also be associated with altered brain physiology.

To the Hindu, enlightenment is union with the absolute. The popular term that describes this experience, yoga, is derived from the Sanskrit root *yuj,* meaning union. Yoga also means the way or vehicle to achieve this union. There are any number of vehicles, or yogas.

Prayer to an unseen but omnipotent God was one of the earliest ways ancient people sought to escape from the mundane life. The ancients also used a variety of drugs to enhance their religious ceremonies. Marijuana has a long association with religion in many parts of the world, especially in the Indian subcontinent. The drug does appear to distort time perception, and it produces a unique state of euphoria that is ineffable. Although at one time drugs played a dominant role in Indian religious practices, in more recent years they have fallen into disfavor. The present-day Indians recognize mainly four yogas: devotion, knowledge, selfless activity, and meditation.

The first yoga, devotion or worship, has a firm association with art. Many places of worship are adorned with exquisite art. Singing is an important component of religious ceremonies all over the world. In some parts of the world, including India, dancing also has ties to religion. In the brain, the holistic nondominant hemisphere is generally acknowledged as the seat of artistic expression and enjoyment.

Concerning the second yoga, the term "knowledge" has a rather limited connotation, at least in the West. According to Indians, there are two types of knowledge: Factual knowledge is the inferior kind; intuitive knowledge is superior. Jesus, Muhammad, and the Buddha had knowledge, but it was a different type from that of Darwin, Newton, and Einstein. Most neuroscientists argue that intuitive knowledge is also a nondominant hemispheric function.

The third yoga, selfless activity, is emphasized by all religions. Selfishness is disdained and selflessness is celebrated. Although there are no established findings on the brain mechanisms responsible for feelings of self, the dominant hemisphere that separates and divides is likely to generate a concept of self, isolated from and independent of the surroundings. The holistic nondominant hemisphere, on the other hand, is likely to blur distinctions between the self and the surroundings. Thus, the first three yogas, like drugs in ceremonial use, appear to enhance the nondominant hemispheric activities.

The fourth yoga, meditation, involves rejection of the mundane world and requires an internal focus. A number of experimental studies, especially studies with the electroencephalogram (EEG), bear this out. We can see the human brain as a map of our evolutionary past. Experimental evidence has revealed components from our neomammalian, paleomammalian, and reptilian ancestry in the human brain. It is highly likely that vestiges from even earlier stages of evolution are present in the depths of our brains. Meditation takes us back past these evolutionary steps, all the way to the very beginning. The creator is present in the creations; the human brain is no exception. The deeper we go, the purer the experience becomes. In a profound meditative trance, subject-object distinctions disappear and time and space cease to exist. Bliss, unsullied by any sort of pain, brightens the experience.

Looking at any of these four yogas shows that they are associated with the right, or nondominant, hemisphere of the brain. In right-handed people, the left hemisphere is the dominant one. The dominant left hemisphere utilizes a linear modus operandi, and it has the tendency to take things apart. Language, mathematics, and logic are some of its functions. The nondominant right hemisphere, on the other hand, is holistic in its function. Enjoyment of art, music, and nature are its gifts. The association between these vehicles for seeking yoga and the nondominant hemisphere points to an association between this part of the brain and religious experiences.

The external world of objects is dependent on perception, but the internal subjective world is not. Research looks at the fascinating question of how the brain experiences the inner subjective world. Whether a meditative trance is neurally mediated is a very difficult question to answer. The brain is derived from matter, and matter is derived from sheer energy. The key question is whether our experiences can take us past the brain and its precursor, matter, to the primordial absolute.

Illogical Logic

Mother's Prescience

It was ebony black outside. A faint orange glimmer of light appeared and shimmered into a line of molten gold. The line elongated to illuminate the pale blue sky above and the woolly white cloud carpet below. Above the canopy of clouds, from the airplane window, sunrises and sunsets are exquisitely beautiful. I could easily get lost in the beauty and serenity even when surrounded by despair and depression. It was early in the morning, and after hearing the news of my mother's death, I was on my way to India to be with my family.

I had arrived at Duke University Medical Center as a professor of psychiatry only a few days before. I was in my new office busily unpacking books when the telephone rang. I picked up the phone, but there was no response. For a second I thought the line was dead. I said "Hello" again. After a pause, my wife responded with a weak "Hello," but nothing else. She started to cry. Bewildered, I asked, "What's the matter, Laura?" "Your mother died," she said. "Your aunt just called." I knew that my mother had suffered a heart attack about two weeks earlier and had been admitted to the University Hospital in Trivandrum, my hometown in India, but I was assured that she was going to recover. I had already made travel plans to go to Trivandrum the next week to see her.

Although as a physician I was aware of the risk of death associated with a heart attack, I had not actually conceptualized the possibility of my mother's death. As a psychiatrist, I had read a great deal about stages of bereavement and dealing with death. As I tell my students, loss of somebody one loves intimately creates a yawning abyss. One's sense of self normally extends beyond one's physical being to include the near and the dear. Relatives, friends, pets, and even lifeless objects form integral parts of one's psyche. Disappearance of such a component, especially when it is sudden, causes mental pain, the same as the physical pain associated with the loss of a limb.

Over the years I had helped many a patient deal with losses. Yet I was helpless in my own despair. Logic and reason, which formed the basis for my training as a scientist, were of no use to me. In desperation I found myself looking and searching everywhere, even in discarded and neglected places. Dust-laden memories stowed away in the dimly lit recesses of my mind were dredged up.

I remembered a trip as a child to the Blue Mountains, which skirt the eastern part of our state in India. My father, a lawyer, was on the board of directors of a tea-manufacturing company, and we were privileged to use their guest house, located in the midst of their tea estates. The bungalow and the tea estate still carried the name of Ashleigh, the original English owners. After Indian independence, most English tea planters left, or were forced to leave, but most plantations still kept their original names. Ms. Ashleigh, the last owner of the tea plantation, built the bungalow, which served as a guest house after her loss of ownership and departure. The bungalow occupied a spot that commanded an exquisite, panoramic view of the valley and the surrounding mountains. The white fence in front of the bungalow stood in sharp contrast to the surrounding dark green hills and was visible from the valley far below.

For children, the tea estate afforded countless opportunities for games and exploration, but to our great surprise, Mother imposed severe restrictions on our activities. My brothers and I considered this completely unfair. Mother said that there were poisonous snakes in the area, and the local gardener, to our dismay, confirmed it. Father, to whom we appealed, for undisclosed reasons chose not to intervene. Several days of severe restrictions had elapsed when news came that a child of our age, the son of an intimate friend of the family, had suddenly died. It was not until then that my mother told us that she had dreamed of kneeling in front of a small coffin and was afraid that this portended one of her own children's death, thus her efforts to protect us from harm.

Mother had many such prescient dreams. Shortly after their marriage, my parents came to Trivandrum, where my father set up his new law practice. Houses within their limited budget were hard to find. Finally, they located a small affordable house, which was not in great demand as it was rumored the house was haunted. Father did not believe in such nonsense, and their financial situation did not grant many choices. Shortly after they moved in, Mother started having a peculiar dream. She would see a young Hindu woman near a chest of drawers in the bedroom, staring back at them. After dreaming the same dream several times, Mother had a clear and vivid image of the woman's facial features. Although she was not greatly bothered by the experience, she told Father about it. He dismissed it out of hand. A few weeks later, when the landlord dropped in, my father casually told him about my mother's dream. The landlord became very excited, almost to the point of agitation. He went immediately to his house, returning with his family album. He wanted to know if Mother could identify the woman in her dream from the album. Mother flipped through the pages and without any hesitation pointed at a picture of a young woman. The landlord identified her as his sister, who had died following childbirth in that very bedroom, several years before.

On another occasion, Mother told Father that a distant relative whom he had not seen in several years would be visiting that evening when they returned from their evening walk. She had had a dream about this the previous night. That evening when they were on their way back from their walk, they saw the relative waiting for them in front of the house.

During the night before her college chemistry practical exam, Mother dreamed that the chemicals she would have were mercury and barium. She also dreamed that in between experiments she ran into her best friend, Eva, in the corridor. Eva was frustrated because she was totally unfamiliar with the test, but Mother pointed out to her that the experiment had been taught when Eva was absent. What my mother dreamed came true the next day, complete with all the fine details.

As a child I had no difficulty in accepting these precognitive dreams; however, as I grew older and "wiser," I began to have difficulties. After finishing my medical training and going to England, my difficulties increased. My training in psychiatry contravened my belief in supernatural or paranormal phenomena. The concept of reality was firmly entrenched in perception, with reason as its exclusive basis.

Mother, however, had had these prescient dreams since she was a child. My father, who rejected all supernatural hocus-pocus with contempt, had developed a healthy respect for my mother's precognitive dreams. I had no doubt whatsoever about the veracity of my mother's experiences.

Yet as a researcher who had spent fifteen years on the study of the brain, I could not overlook the reality of her death. Her sensory faculties, including vision, hearing, taste, smell, and touch, were no longer there. She could not think, she could not reason, and she did not have access to her memory stores. To the scientific mind, life is synonymous with consciousness. Consciousness is dependent upon neural processes that cease to function at the time of death. If one were to believe in an existence beyond the neural process, then one would seriously have to consider the possibility that life is in excess of neural processes and the brain. Mother's precognitive abilities, of which I had no question, were beyond neural processes we know of. Did that mean she, too, had an existence independent of her brain?

Since I'd first left India to go to England for postgraduate studies, every time I left home after vacation, Mother would weep. I knew her thoughts were that this might be the last time she would see me, and the fear that I might not see her again was particularly strong the last time I saw her. I asked her, point-blank, if she was afraid of dying, because I most certainly was. She looked at me curiously and gestured to indicate that she was indifferent to it. She had no question whatsoever about the unseen world and the fallacy of perceiving death as the final end. Her ability to perceive beyond the phenomenal world diminished her fear of death.

The Pan Am flight that took Laura and me from New York to Bombay was tedious. We changed flights both in Frankfurt and Bombay. As the Indian Airline flight to Trivandrum flew over the Western Ghats, the unmistakable landmarks of Kerala became visible. As I looked down, swaying coconut palms, glimmering streams and lagoons, and emerald green paddy fields made their appearance. Soon the plane flew past the coastline and turned around to land. Foamy white breakers of the Arabian Ocean, coconut palm groves, and, lastly, the hard concrete of the runway rose to meet the airplane.

The day was bright, warm, and humid. The familiar sea breeze, warm and salty from the near Samkhumukham (conch shell) beach, caressed and consoled us as we deplaned and walked toward the airport. My two brothers, my mother's sister, and her husband were all waiting for us. I had instructed my brothers to proceed with the funeral without waiting for my arrival. (Traditionally, in Kerala, the dead body is kept at home and not in a funeral home, and burial is conducted within twenty-four hours of death.) From the airport, we went straight to the cemetery. Mother was very fond of roses, and my older brother had brought several with him to place on the fresh grave.

When we were driving home from the cemetery, I asked my brothers if Mother had had any unusual experiences that had presaged her own death. It seemed highly likely that she would have had some such experiences, since she was able to foretell many instances of lesser significance. According to my brothers, two weeks before her death, she was having an afternoon nap in a bedroom adjacent to the lounge. She saw her deceased father come through the open doors. She had not dreamed of him since his death several years before. He was dressed in white garb, as usual, and he looked happy. He came to the bedroom where my mother was lying down and said, "Let's go. It's time." Mother woke up, and the experience was so real that she went outside to make sure that there was nobody around. She immediately recognized it as a precognitive dream and became concerned about what was to follow. She contacted her sister, my aunt, that afternoon. She also got in touch with my brother, who lived close by. She told them that some event of a serious nature involving her was to follow. She was not unusually frightened or agitated and went on with her life as usual. She was in very good health. She was physically active. She seldom used the car. She preferred to walk. She did not suffer from any major medical problems and no one in her family we knew of had died suddenly. That same day, she went to the YWCA, where she was active, and attended a prayer meeting. Later that night, severe chest pain woke her up. During the early hours of the morning, she was taken to the local hospital, where myocardial infarction was diagnosed. Her cardiologist attempted to reassure her and the family, but she insisted that her time was up, although she did not share this with the treatment team, confiding only in her sister and my brother. Her physical condition improved and she was discharged from the intensive care unit to a step-down medical ward. Approximately one week after her hospitalization, after her evening meal, she contacted the nurses with complaints of severe breathlessness. Within a few minutes, she was dead.

On Life in Death

My elderly, silver-haired father was waiting for me in front of our house. The steady flame of the bright brass oil lamp in the room where her body had been laid before the funeral looked serene, sad, and peaceful. According to Syrian Christian tradition, the deceased person's soul lingers on for forty days in the manner of Jesus after his death. In remembrance of this belief, a lighted oil lamp, a Hindu practice, is maintained in the deceased person's home for forty days. The house seemed strangely empty without her, and every nook and corner brought forth her memories. Friends and relatives visited to express their condolence and to share in our grief. The parish priest Mother was very close to visited and reminded us of the Christian belief in the resurrection of the dead.

According to the Christian faith in which I was raised, dead people will come back alive on the day of resurrection. God will judge them and dispatch them either to heaven or to hell. Jesus Christ had already paid for the sins of the world, and therefore, baptized Christians will make it eventually to heaven unless they have committed one of the few unpardonable sins. "From the earth you came, to earth you go," said the parish priest. I grew up with this concept, and it had a great deal of sentimental value for me.

It salved the soreness for a while, but my logical side eventually revolted. Dead bodies do not remain intact; they decay and disappear. How can they come back to life at a later point? Is this resurrection limited to humans only and, if so, why? Considering the number of people who have lived and died during the last 100,000 years or so, there is going to be a rather large crowd to be judged. How can an untutored Stone Age brute be judged with the moral and ethical standards of a modern man? How can even two people from the same time and place be compared on a single scale since we do not all have the same constitution and environment? We do not share the same starting or finishing lines. Variation is an incontrovertible fact of science. It forms the basis for evolution. A number of well-established genetic mechanisms—mutation, migration, and drift—bring about wide-flung variations within species. We have little control over our genetic makeup and, therefore, in all fairness, cannot be held responsible for many of our follies and foibles. All men are created anything but equal.

Another quote came back to me from the time India was mourning the death of Jawaharlal Nehru, its first and most beloved prime minister. A leading newspaper had published a passage from Bhagavad Gītā, the Indian equivalent of the Christian Bible (composed somewhere between 400 B.C. and A.D. 200):

> *He who thinks that this slays and he who thinks that this is slaying; both of them fail to perceive the truth; this one neither slays nor is slain. He is never born, nor does he die at any time, nor having once come to be will he again cease to be. He is unborn, eternal, permanent and primeval. He is not slain when the body is*

> *slain. Weapons do not cleave him, fire does not burn him; waters do not wet him; wind does not dry him . . . He is eternal, all pervading, unchanging and immovable . . . Do not grieve for the death of that which is beyond death. (2.19, 20, 24, 30)*

The passage was not incompatible with available scientific knowledge. Dead bodies disintegrate into more basic elements. Even the process of decomposition nurtures microorganisms. The end products may remain in the soil or be consumed by a microorganism or a plant. The plant in turn may nourish an animal, which, when it dies, again decomposes into basic elements. The cycle of life and death and life goes on ceaselessly in interminable succession. The inorganic becomes life-giving organic that cycles back to inorganic. The substance of which humans are made may one day reappear as a plant or an animal. In that sense, there can be no destruction, only a process of transformation. If there is an end, it is only of the form, not of the substance.

Nothing can be created de novo, and nothing can be expunged out of existence. A pot of water may be boiled dry. Although water is gone from the pot, water exists as steam, every drop of it. A glass may be broken into a thousand shards, but the glass still exists as glass fragments. This concept, in a more basic form, is known to physicists as the first law of thermodynamics—that energy cannot be created or destroyed. It may transmute from one form to another. Alteration is confined to external qualities but not to the basic substance, namely, energy itself.

Raw energy is not perceived; only the qualities that characterize the object it constitutes are perceived. When we examine a flower, we can appreciate its shape, size, texture, fragrance, color, and taste, if it has any. All these are qualities of the flower and not matter of which the flower is made. Thus, we define objects by virtue of their qualities and not their substance. This idea was first articulated by Nāgārjuna, a Buddhist of the Madhyāmaka school, around A.D. 200.

In our time, since Einstein, physicists have been telling us that matter and energy are one and the same and that they are interchangeable. Matter is simply a highly condensed and concentrated form of energy. According to an oft-quoted statement by Einstein, "A human being is part of a whole called the universe, a part limited in time and space. He experiences himself, his thought and feelings as something separate, a kind of optical delusion. This delusion restricts us to our personal desires and affection for a few people nearest to us."[1] The term optical delusion is derived from the psychiatric term "delusional perception," developed by German psychiatrists of Einstein's time. Perception has two components: reception of sensory data and interpretation of those data. In delusional perception, the receiving part is normal but the interpretation mechanism is faulty; for example, a psychiatric patient may correctly identify a bird perched on a nearby tree but may erroneously conclude that the bird has a special message for him.

Several years ago, a psychiatrist under whom I trained in England described death as horrifying. He'd had a near fatal heart attack. He said that when he was confronted with death, it was ghastly and gruesome, like falling into a dark, deep abyss. Since then, I had thought about what it might be like to be dead and always came up with ideas of loneliness, emptiness, and nothingness. I could conceptualize death only in negative terms. The thought that the matter of which one is composed partakes in a never-ending cycle of creation and destruction to some extent expelled the nihilistic thoughts of death as total extinction.

So—Mother was not gone, in a basic sense.

It was comforting to even speculate that she was not trapped in a sensory isolation chamber bereft of her mental faculties until the end of time. She had merely altered her external appearance. She, as we knew her, was there no more, and we certainly missed her. But to know that she had not evaporated into nothing was comforting. She went back to wherever she came from. After death, we go to the same place we were in before birth. This line of thinking, however, raised some obviously intriguing questions about birth and death, and, on a larger scale, about the nature of reality.

On Being Indian

The concept of an unseen reality that underlies the seen world is fundamental to Indian thought. Indians have long recognized the indestructibility of matter and the permanence of the foundational substrate. It might seem strange that Hindu scripture should comfort a Christian who could not find solace in his own religion. Whether that is heresy will depend on what one means by Christianity and Hinduism.

To most people, "Hindu" describes a person who follows a certain Indian religion called Hinduism. This usage, however, is confusing for several reasons. First of all, the word *hind* came from the name of a river, the Sindhu, known as the Indus in English. The terms "Hindu," "Hindustan," and "India" were derived from it. Although the terms "Indian" and "Hindu" have the same etymological roots, the anglicized form, "Indian," denotes someone from the country India, and "Hindu" denotes faith in a certain religion.

As Nehru, the first prime minister of India, said, "Hindi [Hindu] has nothing to do with a religion, and a Muslim or Christian Indian is as much a Hindi [Hindu] as a person who follows Hinduism as a religion."[2]

Hinduism has never existed as an organized religion, in the Western sense. It has never had a centralized pontifical authority and canonical rules, regulations, and rituals. Religious beliefs and practices of the so-called Hindu are so varied, and at times so contradictory, that it is difficult to subsume it all under a single title.

"Hinduism, as a faith," Nehru said, "is vague, amorphous, many-sided, all things to all men. It is hardly possible to define it, or indeed to say definitely whether it is religion or not, in the usual sense of the word."[3]

Mohandas Gandhi, father of independent India, who represented the best of Hinduism and embodied its ancient spirit in full measure, provided the best definition, according to Nehru. "If I were asked to define the Hindu creed, I should simply say: Search after truth through non-violent means. A man may not believe in God and still call himself a Hindu. Hinduism is a relentless pursuit after truth. . . . Hinduism is the religion of truth. Truth is God. Denial of God we have known. Denial of truth we have not known."[4]

According to this definition, a physicist in the intellectual pursuit of the ultimate single equation that combines all known forces of the cosmos will be just as acceptable as a sage who attempts to experience the Absolute via the fevers and ardors of asceticism.

During the course of my life in India, I cannot remember a single attempt at proselytization by any of my many "Hindu" friends. Many of them have come to church with me, both in India and in the United States, and they do not see heresy in it. Their creed has wide-flung boundaries, and it affords them total freedom from dictum and dogma to explore and experiment.

Indians, regardless of their religious beliefs, are philosophically oriented. There are relatively few materialists and pure atheists, although they do exist. Most Indians believe in an afterlife of some sort, and the notion that one's life is affected positively or negatively by one's actions is subscribed to. People of different religions and faiths, Christians included, have deep-rooted respect for Indian philosophy and, to a lesser extent, mythology.

Regardless of the religion a person believes in, philosophy is not a mere intellectual pursuit to the average Indian. It is the warp and woof of existence. According to Radhakrishnan,

> *Philosophy in India is essentially spiritual. It is the intense spirituality of India and not any great political structure or social organization that it has developed that has enabled it to resist the ravages of time and the accidents of history. External invasions and internal dissension came very near crushing its civilization many times in its history. The Greek and the Scythian, the Persian and the Mogul, the French and the English have by turn attempted to suppress it and yet it has its head held high. India has not been finally subdued, and its old flame of spirit is still burning. Throughout its life it has been living with one purpose. It has fought for truth and against error. It may have blundered, but it did what it felt able and called upon to do. The history of*

Indian thought illustrates the endless quest of the mind, ever old, ever new.[5]

Every Indian, from the most to the least sophisticated person, will be at some level familiar with the basic tenets of Indian philosophy, which in its essential form is free of religious dogma. My family was no exception. Although we did not believe in Hindu Gods and did not participate in Hindu rites and rituals, we respected their faith. Mother taught us to honor other people's beliefs. She reminded us that we were Christians mainly because we were born in a Christian family; we had not studied all religions and made an informed choice. There were also other nonintellectual reasons for our respect for even the shallower and cruder versions of Deistic Hinduism.

More than a century ago, when our family moved to the present location, my great-grandfather purchased a house and the surrounding property from a Hindu family. The family were snake worshipers, which meant they simply followed a less-sophisticated version of Hinduism; snake worshiping is not a fundamental or even common component of Hindu life by any stretch of the imagination. The family maintained a snake shrine adjacent to the house, in a part of the property that had never been cleared. It was virtually pitch-dark, with overgrown brush, dangling vines, and large leafy trees that blocked the sunlight. Under the canopy of such a gigantic tree in the mottled light were large numbers of stone snake idols. Hindu women from the family continued to worship at this shrine even after our family purchased the property. The women would visit the snake shrine in wet garments after having a ritual bath in the nearby river. They would leave offerings of eggs and milk for the snakes. Needless to say, the place was infested with a healthy population of snakes, especially cobras.

My great-grandfather, a devout Christian who saw snakes as incarnations of the devil, was singularly unhappy about what was going on. Yet he chose not to intervene, for in India people generally respect others' religious beliefs and do not comment or criticize even when they find others' viewpoints unacceptable. In addition, there is also the hidden fear of the other religions, especially since one is seldom absolutely sure about the impotency of the neighbor's God. An aura of mystery surrounds the age-old Hindu customs, and other religions try to steer clear whenever possible.

After a while, the visits by the women began to flag, and eventually they stopped completely. My great-grandfather, like most Christians, regarded snakes as man's mortal enemy for both practical and religious reasons and proceeded to kill as many snakes as he could find. He laid waste to the snake shrine and smashed the snake idols. While the snake shrine was being annihilated, the priest who officiated at the local temple, approximately two miles away, went into a trance.

During such trances, temple deities are believed to take possession of the priest's body and to communicate through them. These possession states are associated with marked changes in appearance and behavior. Shaking and shivering, frothing and perspiring are common. Those possessed appear to have lost contact with the mundane world. With glassy, detached eyes, they scream and shout, and they sing and dance. Analgesia is common. In some temples in Kerala, possessed priests cut open their scalps to let out rivers of blood, with no external evidence of pain or discomfort. Needless to say, such blood-drenched trance states are awe-inspiring.

Shivering and perspiring, the entranced priest ran from the temple across the paddy fields and over the coconut-palm bridge to our house. My great-grandfather was in front of the house paying the workers their day's wages. My grandfather, a young boy at the time, was by his side. The priest, boiling over with rage, approached my great-grandfather and admonished him in no uncertain terms for killing "my children." "No male children born in this house will survive," he cursed, and ran back to the temple.

At first, the family did not take serious notice of this incident. But, for whatever reason, from then onward, all the boys born in that house died. My great-grandfather was too old to have children, but my grandfather lost two boys and his brother, six. All of them died before the age of ten, of different causes—typhoid, diarrhea, drowning, and so on. Only four girls survived—my mother, her sister, and two cousins. The family finally became so frightened that upon my grandfather's advice, my mother and her sister broke tradition and decided not to go home for childbirth. My mother lost her first child, a boy, of crib death. The other male children they had away from the house survived—my two brothers and I and our four cousins.

The three events mentioned above—namely, the destruction of the snake shrine, the curse, and the death of eight infants—definitely happened. The question is whether there was a causal relationship between the first two and the last. One may argue that the infants' deaths were simply coincidental and not difficult to explain in an underdeveloped country where more than seventy-five years ago accidental and premature deaths were not uncommon. The family, however, did not accept that explanation.

My grandfather's brother, who lost six boys, was a very devout Christian who lived by the book and observed all rules, regulations, rites, and rituals. When his sixth male child was born, after the loss of the previous five, he pleaded with his god to spare his newborn baby. He was too old to have more children. He believed in a living god, kind and benevolent to the devout in desperate need with nowhere else to turn. He spent long periods of time in fasting, praying, and performing other devotional activities. He doted on his only son and virtually kept him under his ever-watchful eye, day and night. The child was not allowed to go anywhere on his own. His son survived until he was eight years of age, when he drowned in a flooded paddy field, during the monsoons. The incident totally crushed my granduncle's belief system and broke him. He no longer had any use for a god who could not spare his child. God either had to be incapable of preventing the tragedy or he did not care. Both were equally unacceptable to my granduncle. Never fully recovering from the deep morass of depression, he vowed not to return to the church, and he discontinued all religious activities. The once-devout Christian became a total atheist and lived out the remainder of his life in solitude. I visited him many times before his death. I would usually find him alone by the nearby brook staring into empty space—a frail old man bent with the burden of broken dreams, laconic and reclusive, sad and vacuous—a feeble shadow of the energetic, sociable person he once was.

Chance, Intuition, and Scientific Thought

This incident and my mother's precognitive dreams are outside the realm of normality and difficult for the logical mind to understand and accept. While there was no question whatsoever that they did happen, to the rational person, the best explanation would be chance.

Although often dismissed lightly, chance occurrences have played, and continue to play, a dominant and important role in all aspects of our lives, at all levels. Chance determines why, when, and where we are born and why, when, and where we die. It influences our genes and environment, which by and large make us into what we are. It also determines many pivotal decisions we make as adults. Even the most dedicated scientist cannot overlook the importance of chance occurrences in the progress of science.

The history of science is replete with landmark discoveries that were chance occurrences. In 1928, while working with a culture of staphylococcus, Alexander Fleming accidentally noticed a bacteria-free circle around a contaminant, a mold growth. Continued research on the mold and the substance it exuded resulted in the discovery of penicillin. In 1895, Wilhelm Conrad Röntgen observed by accident that when an electric current was passed through a discharge tube wrapped in black paper, a nearby fluorescent screen lit up. This incident eventually led to the discovery of x-rays. Ernest Rutherford formulated his nuclear model of the atom in 1911. His student Ernest Marsden, who observed that alpha particles from a radioactive source were deflected when they hit a thin metal foil, and occasionally deflected in unexpected ways, provided the nisus for this momentous finding.

More recently, Sir Roger Penrose, a famous British mathematician, mentioned having had a similar experience. The solution to a mathematical problem he was wrestling with cropped up out of the blue when he was going across a street while engaged in conversation. By the time he got to the other side of the street, the thought disappeared as mysteriously as it had come; however a feeling of elation lingered on. Later on, with considerable effort, Penrose was able to recapture that thought.

Einstein believed that the universe was systematic, symmetrical, and predictable and that caprice and chance had no role in it. "The only unpredictable thing about the universe is that it is predictable," he wrote. On several occasions he declared, "God does not play dice."[6] However, most physicists do not seem to agree. According to Stephen Hawking, "God not only plays dice, but sometimes he throws them where they cannot be seen."[7]

This imprecision found expression in a principle well known to physicists as Heisenberg's uncertainty principle. Talking about subatomic particles that travel in waveforms, Werner Heisenberg concluded that the particles couldn't have a well-defined position and a well-defined velocity at the same time. Stated differently, it is impossible to know the velocity and the location of a particle at the same time.

The unpredictability is exemplified in an imaginary experiment devised by Erwin Schrodinger in 1935. In the experiment, a poison gas and a live cat are both contained in a box made of an opaque material. The bottle of gas is connected to a Geiger counter; Geiger counters are sensitive to radiation. In the experiment, the Geiger counter is exposed to uranium. If a nucleus disintegrates, the consequent radiation will trigger the Geiger counter, which in turn will break the bottle and release the poison gas, killing the cat. The moment of first uranium nucleus disintegration cannot be predicted with any degree of accuracy. Thus, the fate of the cat at any one point of time also cannot be predicted. Statistically, the cat is in limbo between life and death.

Prominent physicists point out to us that chance played a pivotal role in the formation of the sun and the planets—including the planet Earth—from the dregs of the big bang. While the physical forces and elements involved are known, the birth of our galaxy in its existing form will have to be imputed to random occurrences. According to the famous physicist Stephen Hawking, "In fact, a universe like ours with its galaxies and stars is actually quite unlikely. If one would consider the possible constants and laws that would have emerged, the odds against a universe that has produced life like ours are immense."[8]

Naturalists argue that if it were not for some key chance occurrences, humans would never have evolved from the primordial soup. According to the famed anthropologists Richard Leakey and Roger Lewin, "One of the most important developments of evolutionary biology in recent years is that luck, not superiority, plays a cogent role in determining which organisms survive, especially through times of mass extinction. We therefore have to accept that humans are in the company of lucky survivors of cataclysmic convulsions in Earth history, not the modern manifestations of ancient history."[9] They note that our very existence is totally dependent upon chance, considering the two narrow brushes with asteroids in March 1989 and January 1991.

Astronomers tell us that these asteroids, no more than 300 yards in diameter, cruising at a distance from the earth comparable to that of the moon, portend the high likelihood of a future collision of cataclysmic proportions. In fact, in July 1994, comet fragments, some the size of Earth, scarred the face of Jupiter right before our eyes. The birth of the cosmos, the sun, our planet, plant and animal life, and humans were by and large chance occurrences strung together by explicable and plausible phenomena.

Yet we have no clear understanding of this crucial phenomenon "chance." Science dismisses it as inconsequential. For the scientists, it is simply a wastepaper basket into which what we cannot comprehend is tossed.

Scientists still insist that science can explain everything, and that which falls outside of its realm is disdained as superstition, black magic, and witchcraft. As Steven Hawking wrote, "We have thus far been too busy developing theories which describe what the universe is to ask the question, why."[10] Why did humans come into existence? Why are people born, and why do they die? Optimistic proponents of science may argue that eventually science will unravel the answers to all these questions as well. I must confess I am not of their number.

Ancient Indians recognized two sources of sacred knowledge: *śruti,* derived from *śravaṇa* (to hear), and *smṛti,* related to *manāna* (mentation). The former is direct experience, and the latter, interpretation. *Smṛti* is that which is reasoned out and is conditioned by the background and sequence of events and thoughts that preceded it. It is the product of mental efficiency and effort, discussions and debates, comparisons and contrasts. It is based on conclusions drawn by erudite scholars conversant with theology and scripture after much thought. However, it is distinctly human and therefore more subject to error, more open to reification and reinterpretation. Although over the years a number of people have attempted to define and describe *śruti,* the best definition I could find came from Śāṁkara, the preeminent Indian philosopher (to be described later). Śāṁkara, who carries the epithet "Ācārya" (venerable teacher), is noted for his ability to provide clear and lucid explanations for complex philosophical concepts. According to him, *śruti* is not a product of human effort. It blossoms spontaneously and unexpectedly, in the fertile mind. It is a gift and is of ontological significance. Although not wrought by remorseless logic, it is considered superior to factual knowledge. When in conflict, *śruti* supersedes *smṛti. Śruti,* as explained by Śāṁkara, would seem very close but not identical to what is commonly regarded as a serendipitous or chance occurrence.

On a more personal level, in trying to come to terms with my mother's death, logic on its own was insufficient and inadequate. The appeal of the ancient verse from Gītā—the Hindu scripture referred to previously—could not be explained by logic alone; the words stirred the inner recesses of my mind and soul. Although compatible with contemporary scientific thought, the thoughts expressed in the passage did not derive from experiments. They represented the outpourings of an exalted mind from over 2,000 years earlier. Intelligence based on discursive reasoning generates scientific knowledge, which at best is imperfect. Intuitive knowledge valuable even to the scientist is immediate and spontaneous, distinct from discursive or mediate knowledge. Unlike the modern scientist who acknowledges but devalues intuition, the ancient Indian seer gave equal weight and value to both discursive and intuitive knowledge. Intuitive thoughts, spontaneous and inexplicable, were gratefully and reverentially acknowledged and accepted. The exalted outpourings of these minds represented a synthesis of the two strains in contemporary thought—namely, intuition and intellect. This line of approach held more promise for me than any other. Although I was not a Hindu in a sectarian sense, my mother's death prompted me to take a more serious look at Indian philosophy.

The Wellspring of Indian Religious and Philosophical Thought

The roots of Indian culture and civilization upon which Indian philosophy is based may be traced to the civilization that flourished in the fertile flood plains of the river Indus, which now separates India from Pakistan. In the latter half of the nineteenth century, under the hot Indian sun, a group of laborers were toiling building a brick ballast for the railroad track from Lahore to Multan, which is in present-day Pakistan. The brick suppliers knew of a cheap source, a wide expanse of old ruined brick buildings and walls. Although ravaged by time and decay, the kiln-fired bricks were of excellent quality. They pillaged the finely molded bricks, not realizing that they were more than 4,000 years old. To date, for a stretch of some 100 miles, from Lahore to Multan, trains run on tracks resting on a ballast constructed with bricks made before 2500 B.C. Sir Alexander Cunningham, director general of the Archeological Survey of India, quickly recognized the profound historical and cultural significance of this pile of rubble and ruins. Excavations carried out by Cunningham and his successors—Sir John Marshall, Rai Bahadur Daya Ram Sahni, and others—revealed an abundance of small villages from about 3000 B.C., covering a huge area of the Indus alluvial plane. Two capital cities, separated by approximately 350 miles, were identified, and they were dubbed Mohenjo-Daro (Hill of the Dead) in the south and Harappa to the north. The two cities were remarkably uniform in their architectural structure and construction material.

Unlike most ancient cities elsewhere in the world, where roads were enlarged meandering footpaths, the Indus city roads followed a well-planned grid pattern. Private houses utilized the same architectural concepts used in the southern part of India at present. Sleeping quarters, living rooms, storage sites, and so forth enclosed a central courtyard. Indoor plumbing was built into most houses, with the luxury of bathrooms with paved sloping floors. A municipal sewage system drained effluent water from individual houses.

Cattle, buffalo, goats, sheep, pigs, camels, and elephants had all been domesticated, and dog, the old friend, also left its mark in the Indus Valley. The Indus Valley inhabitants utilized copper and stone for household utensils. Articles made from gold, silver, ivory, jade, agate, crystal, and lapis lazuli were also found.

The main occupation was agriculture. Systems of levees and irrigation ditches were constructed to gain control over the water supply. Terra cotta seals, paintings on pottery, and carvings provided a great deal of information on the inhabitants' lifestyle. Toys for the children, ornaments for the women, and tools and weapons for the men were found. Continued research identified artifacts that took the seedlings of Indian civilization deep into the misty realms of the Neolithic age of around 7000 B.C. Bullock carts and boats were used for transportation, the same as in present-day rural India. Although swords, spears, and shields were found, the Indus Valley inhabitants were certainly not a warring people. Craftsmen and artisans produced paintings, ornaments, wood carvings, and figurines of considerable beauty and elegance. Pictographic letters were found on terra cotta seals, though they remain to be deciphered completely.

The beginnings of Indian religious and philosophical thought can be traced to the Indus Valley. Religious relics found in the diggings suggest associations with modern religious practices, especially in South India. Indians worship a variety of animals, including the ape *(hanūmān)*, bull *(nandi)*, eagle *(garuḍa)*, and others; terra cotta seals suggest that Indus Valley inhabitants also worshipped animals. Indus Valley inhabitants worshiped trees, as some Indians still do today. A number of excavated seals and tablets depict what appear to be acacia trees growing inside constructions that resemble temples. Pipal trees and branches of pipal trees are also shown in association with pictures and images of presumed religious significance.

Based on recent findings, the civilization would appear to have taken root around 7000 B.C. and to have reached its heyday in 2500–2000 B.C., after which dissolution and decay began to set in. Invasion by the powerful Āryan nomads, armed with superior weapons and horse-driven chariots from the grasslands of Central Asia, sounded the death knell for the Indus Valley civilization.

Unlike the Indus Valley script, Sanskrit, the language of the Āryans was to remain a spoken language for several centuries until around 400 B.C., when Āryans completely overwhelmed and overshadowed every aspect of the Indus Valley civilization and lifestyle. The advancing Āryan armies chased the Indus Valley inhabitants to the southern parts of the Indian subcontinent. By about 800 B.C., Āryans had learned how to make utensils and weapons from iron ore, which enabled them to surge forward and conquer most of the eastern and southern regions of India. Āryan influence became visible in northern parts of India, while the Dravidian culture, presumably derived from the Indus Valley, held sway in the South. In recent years, the genealogy of Āryans and Dravidians, and their relationship to the Indus Valley inhabitants, have gathered political momentum in India and become fiercely controversial.

Decimation of the civilization that flourished in the Indus Valley sank India into illiteracy for the next 1,000 years or so until Sanskrit evolved into a written form around 400 B.C. Brāhmi, the forerunner of Sanskrit script, was probably derived from the pre-Āryan scripts. According to Suniti Kumar Chatterji, editor of the first volume of *The Cultural Heritage of India,* "The Āryans, probably, had no system of writing of their own . . . After they settled down on the soil of India, a modified form of the late Sind-Punjab script was in all likelihood adopted to write the Āryan language, which was at that time a kind of late Vedic Sanskrit."[11]

The Āryans brought with them a collection of Sanskrit hymns, called the Vedas. Vedas are religious scripts of great antiquity. The term "Veda," derived from the root Sanskrit word *vid* (to know), stands for divine knowledge. The term Veda is loosely applied to other sacred texts as well; for example, the Syrian Christians of South India to which I belong, call the Bible a Veda. Of the four Vedas, Ṛg, Yagjur, Sāma, and Atharva, the first one is considered the best and most important. Ṛg Veda is believed to have been composed around 1500 B.C.

The Vedas are a medley of many things: rites and rituals, pleadings and prayers, sacraments and sacrifices. They extol the virtues of their gods, disparage the shortcomings of their enemies, and express wonder and awe at the mystery of nature. They attempt to comprehend, explain, and interpret life and death, creation and destruction. While a great deal of the material contained has minimal use or relevance to our present day, their murky depths do contain rare, coruscating diamonds. Just as the Old Testament can be read for its poetry, so the Vedas are poetic compilations celebrated by Sanskrit scholars for their literary excellence and beauty.

The Vedas grew in sweep and scope over many centuries, and many contributed to it. The concluding portions of the Veda, (Vedānta) constitute the Upaniṣads. Like Vedas, they were also composed over a stretch of centuries, starting from around 600 B.C.

Upaniṣads are the crown jewels of Indian philosophical thought. In their forest abodes, the disciples sat around the sage while he shared his thoughts and insights with them. The Upaniṣads are respected and revered by Indians as a whole, by all castes, creeds, and religions. They address a wide range of topics from creation to destruction, physics to philosophy, sacred to secular. While some of them are theistic, some others, especially the ten principal Upaniṣads, are free of religious dogma.

Blind belief and dogma are conspicuous by their absence. True knowledge cannot be accessed without purity of intent and disciplined conduct. The Upaniṣadic sages were passionately attached to the concept of freedom. They insisted that inquiry about the truth of things must in no manner be fettered by fear of any sort. "The wise man should cross by the boat of Brahman [the divine], all streams which cause fear" (Śvetāśvatara Upaniṣad, 2, 8).

No account of Indian philosophy and religion is complete without reference to the great Buddha. He looms large among the many personages India produced. No single person changed the course of Indian thought, left such an indelible imprint in Indian philosophy, and revised the form and fabric of social mores in India as the Buddha did. He was the first to challenge the elitist idea that salvation was reserved for the highborn. He replaced rites and rituals with ethics and morality. In his religion, magic and miracles were deemphasized, conduct and lifestyle were given prime significance. Kindness and compassion became central; simplicity and moderation were essential. Excesses in self-indulgence and abnegation were to be eschewed. Reason replaced blind faith and idol worship. Dialectical elements that were evident in pre-Buddhist Upaniṣads and of which he was aware gained new strength and substance.

So profound was his philosophy, so persuasive his reasoning, and so appealing his personality that the religion the Buddha propounded eclipsed Vedic Hinduism for more 1,000 years. Under the emperor Aśoka, it became the official religion of India, and during his reign, it spread to the surrounding Asian countries, where it continues to be dominant today, and beyond. The five precepts of Buddhism (Pañca-śīla) became the most respected code of ethics in India and the Buddhist wheel of Dharma (law of life) adorns the Indian flag. The seedlings he planted burst forth into mighty oaks: the conservative Theravāda, the intellectual Madhyāmaka, and the experiential Vajrāyaṇa, to mention a few. Even the scholastic, Advaita of Śaṁkara, has its roots in Buddhism. The Buddha's breadth of vision and depth of insight remain unrivaled.

Vardhamāna (599–527 B.C.), a contemporary of the Buddha, expounded Jainism, a religion with a number of similarities to Buddhism. Like Buddhism, it was indifferent to the Vedas, rejected the Vedic pantheon, and reprobated the caste hierarchy the Āryans established. Injury to all life forms, including the smallest insect, was anathema to the Jainas, and Jainism placed greater emphasis on self-denial and asceticism.

Theistic thought became stronger and more visible during the next phase, that is, the epic Purānic period from 500 B.C. to A.D.

500. This period also saw the fusion of Āryan and Dravidian religions (probably derivations of the Indus Valley religions). This tendency is seen most clearly in the two epics, Rāmāyaṇa and Mahābhārata.

Rāmāyaṇa, which preceded Mahābhārata, is the story of King Rāma, who is presented as the exemplar of virtue and perfection. Rāma sought the help of the aboriginal tribes of the South in retrieving his wife, Sīta, abducted by the anti-god Rāvaṇa, who resided in the present-day Sri Lanka.

Mahābhārata is the longest epic poem in the world, with over 100,000 verses arranged in 2,009 chapters. It is eight times the size of the Iliad and the Odyssey combined. Mahābhārata is the tale of the great war waged between two branches of the Bhāratas royal family. Some authors, like the famous Indian philosopher Radhakrishnan, hold that the original event was probably a non-Āryan one, although in later years, it was Āryanized. The widely acknowledged date of the great Kurukṣtra war of Mahābhārata is before 3000 B.C. The well-known Hindu God Kṛṣṇa (associated with the Hare-Krishna movement in the West) made his appearance in this epic. The dark-skinned Kṛṣṇa is often regarded as a Dravidian king who was deified in the epics. The much earlier Ṛg Veda spoke of a non-Āryan chief, Kṛṣṇa, who fought the Āryan godhead Indra on the banks of Amsumati.

The Purāṇas are a complex corpus of mythology, cosmology, law codes, ritualism, and devotionalism. In many ways, they resemble Greek mythology, and they have tremendous appeal to the masses in India.

From the early days, science *(śāstra),* which received a great deal of attention, did not contradict and compete with the Vedas; in fact, the precursors of modern science were regarded as appendages of the Vedas. Even the later versions of *śāstras* may be called sciences only in a very broad sense, as they also included subjects that in contemporary terminology are subsumed under art. Here, *arthaśāstra* (economics), *rāsaśāstra* (chemistry), and *jyotiśāstra* (astrology) can be found side by side with *nādyaśāstra* (drama), and *nṛttaśāstra* (dance). During the same period, *arthaśāstra* of Kautilya provided weights and measures.

It is not widely known in the West that numerals were first developed in India and that it was the Arabs who took them to the rest of the world. Nor is it common knowledge that the use of alphabetic letters in mathematical equations (algebra) had their origin in India. Āryabhaṭa (A.D. 476–550) discovered an accurate approximation for pi (3.1416) and introduced the versed sine function (1 minus the cosine of an angle). He came up with the concept of zero and decimals. He also declared that the earth rotated on its axis, giving the illusion of the circular movements of the sky. His textbook on mathematics, Āryabhāṭiya, dealt with algebra, trigonometry, and arithmetic. Varāhamihira (A.D. 505–587) produced Pañcasiddāntika (Five Treatises), a compendium of Greek, Egyptian, Roman, and Indian astronomy of the time.

Taxila, or Takshaśila, a university in the northwestern part of India, predated the Buddha (567–487 B.C.), whose physician, Jeevaka Komārabhacha, was a Taxila medical graduate. According to Buddhist scripture (Vinayapiṭaka, Mahāvagga, VIII, 6, 7), after seven years of rigorous training, Jeevaka Komārabhacha asked his teacher when he would be fully qualified. The teacher gave him a shovel and asked him to find a plant with no medicinal value. After much searching, the student came back exhausted and frustrated; he could not find such a plant. The teacher assured him that his training was complete and he was ready to practice Ayurveda.

Pānini, the fourth century B.C. Sanskrit grammarian, was also trained in Taxila. Nalanda, another ancient Buddhist university that came into prominence later (second century B.C. to A.D. eighth century), accommodated several thousand students and over 1,500 teachers. Both universities attracted students from all over India, as well as overseas, including Central Asia and Afghanistan in the West and China in the East. There was a close relationship between Greece and India, exemplified by the Sanskrit translation Yavanajātaka (Greek Astrology) by Sphujidhvaja (A.D. 269–270).

Ancient Indian astrologers and mathematicians were able to describe planetary behavior with a considerable degree of accuracy. They were able to map out the zodiacal belt, follow the motions of the moon and the sun, determine the moon's synodical revolution, and divide the ecliptic into twenty-seven and twenty-eight parts.

They were particularly conversant with Jupiter (Brahaspati), on which basis they computed two chronological cycles; one of solar twelve-year duration based on a single revolution of the planet, and a second one of five Jupiter years (sixty solar years). The oldest version of the modern calendar, based on a solar year of 360 days divided into twelve lunar months of twenty-eight days, can be found in Atharva Veda (1000 B.C.).

Most of the ancient cultures, including those of Greeks, Romans, Persians, and Arabs, did not have numbers above 1,000, and certainly not above 10,000. Evidence that Indians were able to deal with much larger numbers can be seen even in *śāstras* of inequity. For example, Dharmaśāstra of Manu, composed around 400 B.C., and the Purāṇas provided a detailed description of the cosmic cycle. The system was subsequently modified by the mathematicians Āryabhaṭa, Varāhamihira, and Brahmagupta (A.D. seventh century). The cosmic cycle comprises a day and night of Brahman (the root Sanskrit word *brh* means to grow, burst forth), each of which is 4.32 billion years. One day and one night of Brahman is each made up of 1,000 Manvantaras, one of which lasts 4,320,000 sidereal years. Each Manvantara (also called Mahāyuga) is further broken down into four yugas of varying lengths. (According to Āryabhaṭa's calculations, they are of equal duration—1,080,000 years).

The first yuga was named Satya (Truth) when the universe was pure. Compatible with the modern concept of entropy, increasing dissolution and decay marked the succeeding yugas, Dwipāra and Treta. We unfortunately occupy the last and worst one, Kali, which started with the first recorded war in 3102 B.C.

In his well-known book *Cosmos,* Carl Sagan states, "The Hindu religion is the only one of the world's great faiths dedicated to the idea that the cosmos itself undergoes an immense, indeed infinite number of deaths and rebirths. It is the only religion in which time scales correspond, no doubt by accident, to those of modern scientific cosmology. Its cycles run from our ordinary day and night to a day and night of Brahman, 8.64 billion years long . . . about half time since the Big Bang."[12]

It is indeed remarkable that the ancient Indian, by accident, was able to pick the correct number from the grand total of 8.64 billion, if not infinite, numbers. It is also of interest that Āryabhaṭa, the same mathematician who developed the concept of zero, the decimal system, and the approximation for pi and who also wrote the most complete textbook on mathematics of the time, was instrumental in picking the number "by accident."

Significant progress was also made in medicine, with two notable contributions that came into being during the early centuries of the Common Era, *caraka samhita* (medicine) by Caraka and *śuśruta samhita* (surgery) by Śuśruta. Caraka's text, several times the size of the book by Hippocrates, gives a detailed classification of diseases based on signs, symptoms, and their management. Śuśruta provides explicit descriptions of a variety of surgical techniques, including cesarean section, cataract removal, and limb amputations, and the surgical instruments needed.

The well-known eight basic notes in the Western musical scale were also developed in ancient India. They were derived from Sāma Veda, the melodious rendition of Ṛg Veda verses, composed some time between 1000 B.C. and 500 B.C. The tones employed by Sāma Veda formed the basis for subsequent development of popular and classical music in India and elsewhere. The Persians learned the system, and once again, the Arabs took it from Persia to Europe in the eleventh century.

The material I came across was remote in time but not in thought. The ancient Indians were more intent on seeking unity than separation. In general, the Indian mind is more synthetic, unlike the classic scientific one, which is more analytical. The pursuit after that which unites and binds is evident throughout the history of Indian philosophical thought. Ancient Indian philosophy shows that science and art, intellect and intuition, and subjective and objective sources of information can all be fused into a single, meaningful whole.

I did not feel that the numerical logic and inductive reasoning that characterize contemporary science alone could unravel the phenomena that caught my attention: spiritual awakening, paranormal phenomena, life, death, and so on. However, I was not interested in blind faith and mindless acceptance. Although intuitive experiences may be of cardinal significance, one does not need to abandon reason and logic. Intuition can be and has been supported by intellect, and vice versa.

Although much is known about the physiology of birth and death, the topic of interest here seemed to belong to another realm. The path to that realm may wend its way through the metric pattern of scientific thinking, but for me, it had to go beyond. One can learn a great deal about the violin by studying its size and shape and the wood and metal of which it is made. But to really know the violin, one has to listen to it. The mindset that sustains and supports belief in and yearning for the "beyond" is called spirituality.

Seeing Is Not Believing

Spirituality and Materialism

Indians are fond of the Buddhist story (Udāna vi. 4:66–69) of the blind men of "Indostan" and the elephant. Based on it, John Godfrey Saxe composed a poem in English. Six blind men who had heard a great deal about the elephant wanted to have a better feel for it. They ran their hands over a tame elephant. The first one, who touched its flank, concluded that the elephant was like a wall. The second man happened to feel its tusk, and he insisted that the elephant was like a spear. Feeling its squirming trunk, the third declared that the beast was like a snake. The fourth reached out and touched the animal's leg and stated that the animal resembled a tree. The fifth, who inspected its ear, was of the opinion that the elephant was identical to a fan. The last, after getting hold of the swinging tail, was convinced that the animal was like a rope. As Saxe says at the end of his story,

> *And so these men of Indostan*
> *Disputed loud and long,*
> *Each in his own opinion*
> *Exceeding stiff and strong,*
> *Though each was partly in the right*
> *And all were in the wrong!*[1]

The concept of spirituality is very similar to the blind men's descriptions of the elephant. An extensive body of literature is available on spirituality, reflecting diverse and wide-ranging approaches to the concept. In my study of the subject, I came across some very religious people to whom spirituality was closely tied to their faith. They insist that true spirituality has their faith as its basis; all else is worthless dross. Devotees who are less rigid allow a spiritual basis for other religions as well, although many regard theirs as the purest. Alcoholics Anonymous argues that spirituality is totally independent of religion and can be supported and sustained by belief in a higher power. The Godhead is dispensable. Several artists I came across consider art and the aesthetic experience of it highly spiritual, and to them, it has nothing whatsoever to do with a higher power and God. To the intellectual highbrows, spirituality is an abstract philosophical concept that explains everything: the cosmos, life, and living.

The *Encyclopedia Britannica* (1998) provides a comprehensive definition and description. Spiritualism is a characteristic of any system of thought that affirms the existence of immaterial reality, imperceptible to the senses. Thus defined, spiritualism embraces a vast array of highly diversified philosophical views. Most clearly, it applies to any philosophy accepting the notion of an infinite, personal God, the immortality of the soul, or the immateriality of the intellect and will. Less obviously, it includes belief in such ideas as finite cosmic forces or a universal mind, provided that they transcend the limits of gross materialistic interpretation.

According to the ancient Chinese, the phenomenal world is made up of the opposites Yin and Yang, which literally mean "dark side and sunny side." The concept appears for the first time in the *Hsi tz'u,* or Appended Explanations, in the fourth century B.C.: "One time Yin and one time Yang. This is the Tao" (*Encyclopedia Britannica,* 1998). Yin and Yang represent two intertwining, mutually complementary, interdependent polarities of the same principle—male and female, good and bad, hot and cold, light and darkness, and so forth. Each phrase carries elements of the opposite within. Together, Yin and Yang symbolizes the intercalation of opposites in the universe. Yin and Yang slowly became a foundational concept in Chinese and, later, Japanese philosophies. In ancient China, an entire school of philosophy was founded on its basis, and it influenced all aspects of Chinese science, and art as well. In Japan, it became In-yo, permeating Japanese life at every level, and it has stood the test of time. More recently, Westerners—even scientists—have adopted it.

Since the world is made of opposites, the best and perhaps the only way to understand any linguistic concept is by comparison and contrast with its opposite. In the absence of darkness, it will be very difficult to conceptualize light. Heat is the absence of cold, or vice versa.

Materialism, according to the encyclopedia, is the polar opposite of spirituality. Materialism, as the term implies, is belief in the doctrine that matter is the ultimate reality. According to this view, the cosmos and all related phenomena can be explained on the basis of physical sciences. Materialists are strongly opposed to religion and theology. They revile and reject priest craft and all forms of what they regard as sorcery and superstition. They accept only the authority of physical proof and perceptual evidence. They have no time for or interest in the nonexistent before and after, unseen heaven and hell, and the fantasized devil and divine.

Both concepts, materialism and spirituality, are ancient, perhaps as ancient as humans. Materialistic tradition in Western philosophy began with the fifth century B.C. Greek philosophers Leucippus and Democritus. The latter insisted that the world is made up exclusively of atoms, which cannot be further subdivided. All versions in appearances are due to permutations of atoms. The tradition was carried on by a number of later philosophers. The great French philosopher René Descartes epitomized modern materialism, but he also accepted the ethereal nature of the human mind. He advanced a dualistic philosophy that cleaved mind from matter.

The spiritual tradition in the West commenced with the fifth century B.C. Greek philosopher Pindar. He attributed a divine origin to the soul, which is temporarily housed in the body, only to return to the source for its dessert of reward or punishment after death. Both Aristotle and his disciple Plato may be regarded as supporters of the spiritual tradition, since they recognized the soul and saw God as the ultimate actuality.

Concepts of spirituality and materialism in India precede 1500 B.C. Although the dominant theme in Indian philosophy has always been spiritualistic, its entire history is spotted with materialism. Even some hymns of Ṛg Veda contain materialistic elements.

Early Buddhist scriptures also refer to this doctrine: "Man is composed of four elements. When man dies, the earthly element returns and relapses into the earth. The watery element returns into the water, the fiery element returns into the fire, and the airy elements return into the air; the senses pass into space. Wise and fool alike, when the body perishes, do not exist any longer."[2]

Early Buddhism definitely denied the existence of the soul and God. Buddhists argued that both the internal and external world are imagined and impermanent. The Buddha inveighed against sacraments and ceremonies, religious institutions and priesthood. On the other hand, Buddhism departed from classical materialism by accepting the doctrine of reincarnation, enlightenment *(nirvāṇa), dhammas* (the cosmic code, with special emphasis on conduct), and firm rejection of the verity of perception. True reality, according to the Buddhists, is beyond the sentient world—they call it *śūnyatā,* or nothingness (in a perceptual sense).

Schools of thought much more materialistic than Buddhism were in existence in ancient India, but at present very little is known about them. They are collectively called Lokāyata or Cārvāka. The proponents held Lokāyata was the only truth; in it, only perceptual evidence was authority. Earth, water, fire, and air constituted the basic elements. Enjoyment of sensual pleasures was the only purpose of human existence. Matter can think. There is no other world, besides or beyond the present one. Thus, one should have a good time so long as life lasts, for death is the end of all. There is no incorporeal judge or final judgment, no heavenly reward for good conduct or hellish punishment for bad ones. Mortals created concepts of good and bad for their convenience, mostly, but not exclusively, to facilitate and support smooth and harmonious community living. Supernatural notions in general and, in particular, those related to sin and punishments, were concocted by a few to manipulate and control the masses.

Although mainstream Indian philosophy was remarkable for its respect for freedom to think and act, extremists eager and willing to drown and destroy dissenting voices were certainly present. Much of the ancient literature on materialism in India was probably destroyed by the ecclesiasts, aiming to establish their monopoly.

Materialism, according to the encyclopedia, is the view that all facts, including those about the human mind, are causally dependent upon or even reducible to physical processes. To be more specific, matter is ultimate reality. The universe and life can be fully and completely explained on the basis of physical sciences. Life ceases to exist after death, and the ultimate goal in life is enjoyment of pleasures of the senses.

From the foregoing discussion, the close link between perception and matter should be obvious. Matter can be recognized only through the perceptual apparatus. So long as matter is in the objective space, knowledge about it will have to depend upon perception. Matter cannot be known directly. Only signals emanating from matter can be perceived; for example, an object is seen exclusively from the light reflecting off it, and the eddies in air it produces are heard. Any discussion of the objective space and matter therein will therefore have to take into account the perceptual process and its limitations. Thus, the materialistic dictum that matter is reality should be extended to include perception, since it is the only way matter can be known.

Perception and the Human Brain

The human brain is a nexus of over 100 billion neurons. All types of brain activity, from most simple to most complex, depend upon efficient interactions between groups of neurons. During the execution of any one activity, some neurons are stimulated, while others are silenced. Still others remain in their perpetual state of readiness. The brain is constantly humming with activity, like a beehive.

Efficiency of brain function is thus centered on an effective system of communication between neuronal populations, which turn on and off depending upon the demands of the task on hand. The brain, or at least some part of it, is perpetually active so long as there is life. The combined electrical activity of the brain may be recorded using pairs of electrodes applied externally to the scalp. Undulations in the potential differences between each pair of electrodes produce peaks and troughs on a line graph. Known as electroencephalography, this technique was first developed by Hans Berger in 1929.

Communication between neurons depends upon transfer of electrical pulses known as "action potentials." Synapses are copula between neurons; electrical and chemical subtypes of synapses also exist. The action potential flows passively across electrical synapses. Chemical synapses, which constitute the majority, rely on chemicals called neurotransmitters for carrying the action potential from one neuron to the next (Figure 2.1). The brain utilizes a wide variety of chemicals as transmitters. While some neurotransmitters stimulate the receiving neuron (for example, glutamate), others inhibit it (glycine).

Although the brain is responsible for all types of perception, it is incapable of appreciating unprocessed, raw sensory stimuli. It is insensitive to light, sound, touch, taste, smell, and even pain. Neurosurgery may be performed without anesthetizing the brain. The surgically exposed brain cannot detect shining lights, noises, touching, or even cutting. The brain is an electrical organ. It recognizes only electrical impulses and chemical changes that trigger electric currents. Sensory organs convert signals from the environment into electrical pulses, and nerves carry them to the brain.

Skin contains a variety of such transducers called receptors, which are usually divided into mechanical receptors (touch, pressure, and so on), nociceptors (pain), and thermoceptors (heat and cold). The latter two are free nerve endings. Muscle spindles, Golgi tendon organs, and joint receptors provide information on position and tension. All sensory receptors convert sensory stimuli into action potentials.

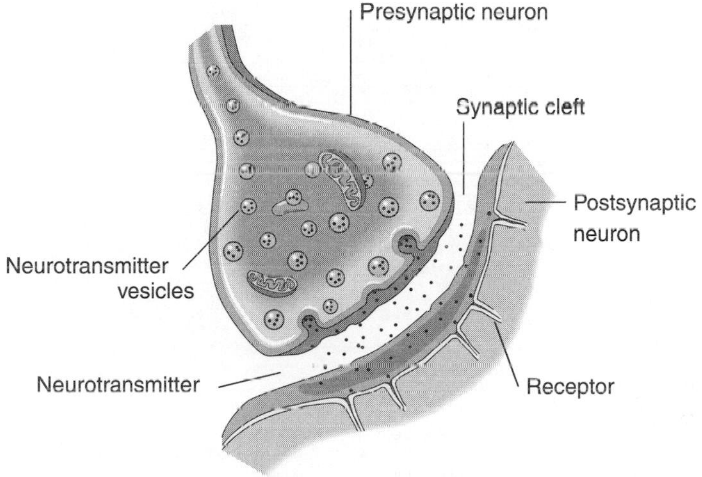

FIGURE 2.1 ***Synapse.*** *Upon the arrival of a nerve impulse, the presynaptic neuron releases neurotransmitters into the synaptic cleft. The neurotransmitter will diffuse across and bind to receptors on the postsynaptic membrane.*

Action potentials ascend through the spinal cord nerve tracks upward to the brain. Upon entering the brain, they converge toward a relay station called the thalamus (diencephalon), whence nerve fibers go to the sensory cortex, on the surface of the brain (Figure 2.2). The amount of cortical and thalamic space allocated to various body parts corresponds to the density of sensory receptors in those regions. The brain is able to appreciate not only the quality and intensity of sensory stimulation but also the location.

The basic principles of the auditory system are similar. Sound is composed of pressure waves generated by vibrating air molecules, similar to ripples in a body of water. However, unlike the two-dimensional water surfaces, sound waves travel in three dimensions. The tympanic membrane inside the ear picks up sound waves and transmits the vibrations to three small interconnected middle ear bones called ossicles. They amplify the vibrations of the tympanic membrane severalfold and turn them over to the cochlea

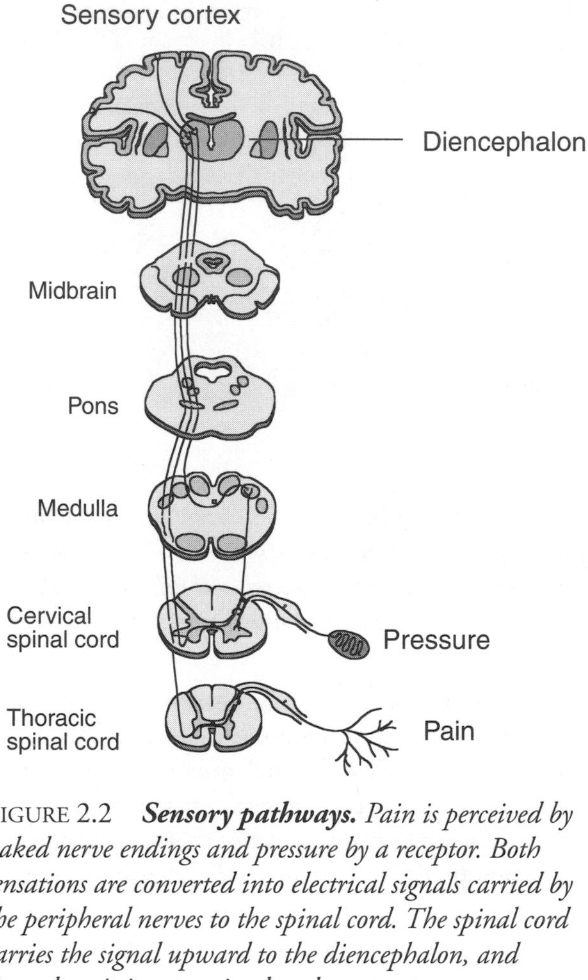

FIGURE 2.2 ***Sensory pathways.*** *Pain is perceived by naked nerve endings and pressure by a receptor. Both sensations are converted into electrical signals carried by the peripheral nerves to the spinal cord. The spinal cord carries the signal upward to the diencephalon, and from there it is transmitted to the sensory cortex.*

of the inner ear. The hair cells of the cochlea serve as the transducers, which convert sound energy into action potential (Figure 2.3). The auditory nerve takes the signal to the brain.

Although the least sensitive in humans, the olfactory system is considered to be the oldest from an evolutionary perspective. The

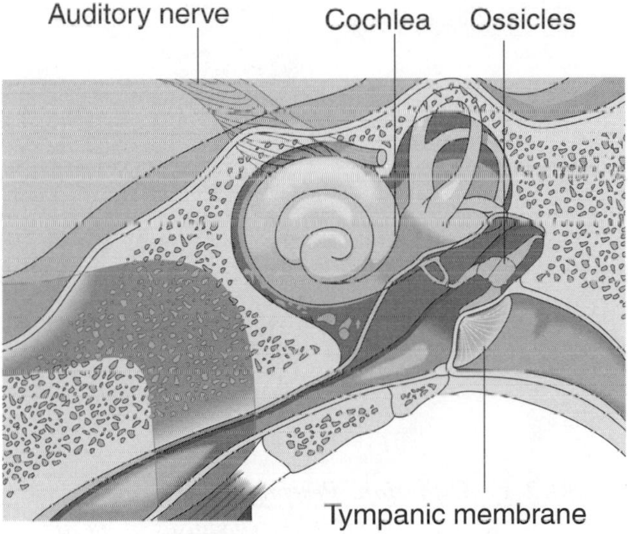

FIGURE 2.3 *Sound waves produce vibration of the tympanic membrane. Ossicles in the middle ear amplify these vibrations and transmit them to the cochlea of the inner ear where the vibrations are converted into electrical signals and carried to the brain.*

mucous membrane, which lines the interior of the nose and is called the olfactory epithelium, serves as the smell transducer. Action potentials thus generated go to the olfactory bulb from where they are transmitted to the brain (Figure 2.4).

Functionally and anatomically, taste perception is close to olfaction. Good food usually has an appealing aroma; malodorous substances are unlikely to be tasty. The tongue and mucous membrane of the oral cavity can detect a variety of tastes, of which four are traditionally considered principal: salt, sweet, sour, and bitter. Humans have approximately 4,000 taste buds distributed throughout the oral cavity, which contain taste cells. The taste cells transduce different tastes into action potentials, which are sent to the brain. The burning taste of chili pepper, for example, is not mediated by the taste

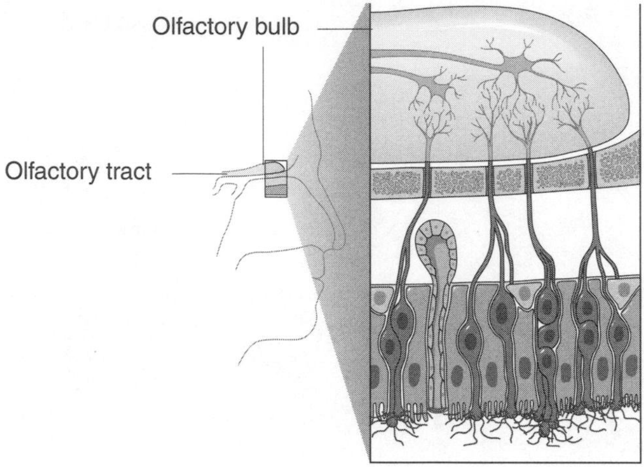

FIGURE 2.4 ***Olfaction.*** *Receptors on the nasal mucous membrane respond to the chemicals in the air and produce action potentials that are carried to the brain by the olfactory bulb and then the olfactory tract.*

bud. The capsaicin present in the pepper stimulates nociceptive receptors. The resultant sensation, though akin to pain, is found desirable by one in three people. This pleasure associated with pain is not a unique characteristic of chili peppers. The same phenomenon can be seen elsewhere in the body as well. A bite inflicted lovingly by a sweetheart might be perceived as highly pleasurable, whereas an identical bite administered by a stranger will be distinctly painful.

In humans, the visual system is the most complex and most advanced. It provides a wide range of information, including the location, size, shape, color, texture, speed of movement, direction of movement, and so forth. The cornea and lens focus images on the retina (Figure 2.5). Developmentally, the retina is an extension of the brain. The light-sensitive cells of the retina, referenced as photoreceptors, are of two types, rods and cones. Rods and cones differ in their shape, the type of photo pigment they contain, their distribution across the retina, and so on. The rods have very low

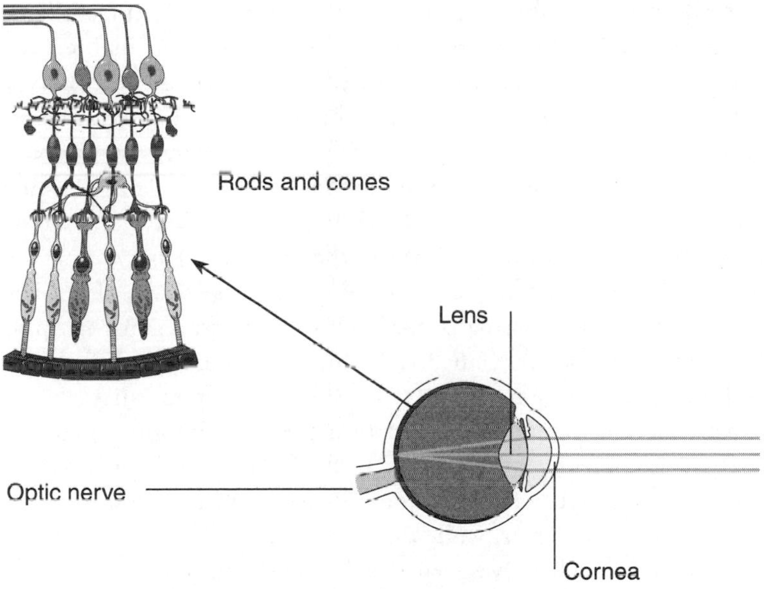

FIGURE 2.5 ***Vision.*** *The cornea and the lens focus light on the retina. Visible light produces a chemical change in the rods and cones of the retina. The action potentials thus produced are carried to the brain by the optic nerves.*

spatial resolution but are extremely light sensitive. Cones, on the other hand, have very high spatial resolution but are relatively insensitive to light. The cones also provide information on color. The primary colors are blue, green, and red. Three sets of cones contain pigments sensitive to the three colors. Light is tracts of energy packets vibrating at different frequencies. The light-sensitive pigment changes upon absorption of light, stimulating action potentials. The ganglion cells of the retina convey the electrical signals through nerve fibers, which exit the retina through a specialized region in the nasal part called the optic disk.

Nerve fibers carrying visual information proceed from the optic disk to the brain via the optic nerve. Further down, optic nerves from each eyeball cross in an X-shaped configuration, so that the fibers derived from the nasal and lateral halves of two retinae (representing visual fields from one side) are carried together (Figure 2.6).

In humans, 60 percent of the fibers cross over in the chiasma, while the other 40 percent do not. The result of the intermingling of nerve fibers from both eyes is the first step in a series of processes to extract three-dimensional vision from the two-dimensional images from each retina. In animals with laterally placed eyes, each of which see entirely different images, the crossing of nerve fibers is more complete than in humans, who have their eyes closer together. From the chiasma, visual nerve fibers travel in the optic track toward the lateral geniculate bodies on either side of the brain and from there move to the nerve cells in the occipital cortex.

Sensory information converted into electrical impulses is transmitted to different parts of the brain for further processing. Sensations from the eye, ear, nose, and tongue reach the brain directly through separate nerves, while those from skin, muscles, and joints travel first through nerves that connect with the spinal cord. The upper end of the spinal cord expands and enlarges to form the main body of the brain. The brain stem is the stalk that connects the spinal cord to the upper parts of the brain (Figure 2.7). Higher up, it is contiguous with the midbrain, which in turn merges with the diencephalon (Figure 2.2).

The diencephalon has two main components: the thalamus, which relays sensory information from the spinal cord to the higher brain center, and the hypothalamus, which mediates autonomic (blood pressure, heart rate, and the like) and neuroendocrine (glandular) functions. The hypothalamus also regulates food intake, water balance, sexual rhythms, and rudimentary emotions such as anger and fear.

The crown canopy, which mushrooms over the brain stem, is the cerebral cortex (Figure 2.8). A central cleft dissects the mushroom into two separate halves (hemispheres), connected at the bottom of the cleavage by a bridge called the corpus callosum. Cerebral hemispheres represent the highest conceptual and sensory, motor, autonomic, and endocrine functions; their mantle of gray matter is made up predominantly of neurons, which have nerve cell path-

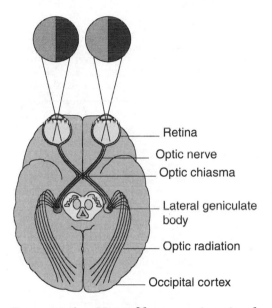

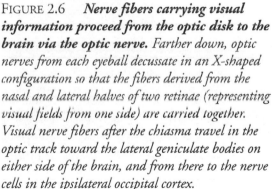

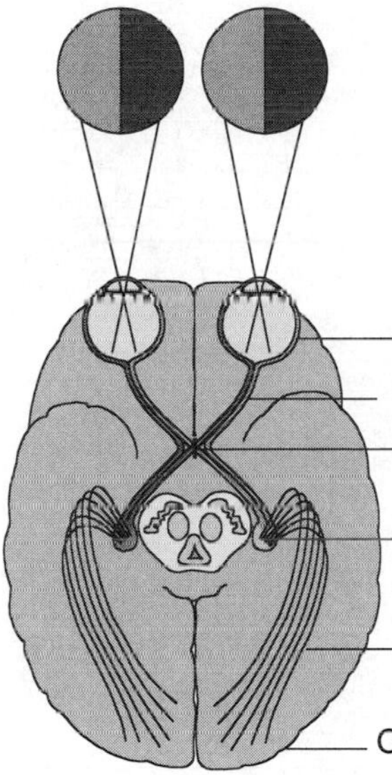

FIGURE 2.6 ***Nerve fibers carrying visual information proceed from the optic disk to the brain via the optic nerve.*** *Farther down, optic nerves from each eyeball decussate in an X-shaped configuration so that the fibers derived from the nasal and lateral halves of two retinae (representing visual fields from one side) are carried together. Visual nerve fibers after the chiasma travel in the optic track toward the lateral geniculate bodies on either side of the brain, and from there to the nerve cells in the ipsilateral occipital cortex.*

ways reaching down into the subcortical and brain stem regions that are used to establish contact with other parts of the brain and the entire body. Large bands of nerve cells are contained within sheaths of myelin—much like insulated electrical wires—and constitute the white matter of the brain. The two cerebral halves are further subdivided into upper and lower halves by another fissure on the lateral surface called the Sylvian sulcus. Other smaller sulci

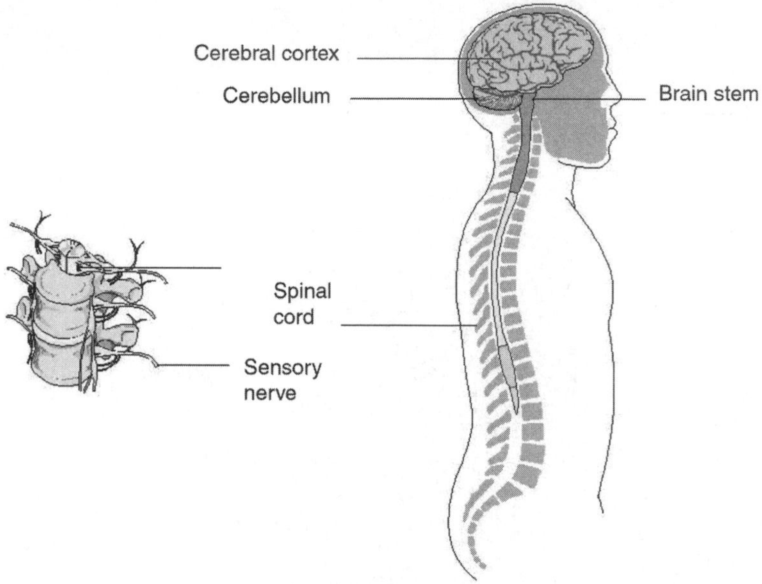

FIGURE 2.7 ***Spinal cord and brain.*** *Peripheral nerves carry sensory information to the spinal cord, which takes it to the brain via the brain stem.*

subdivide the cortical surface into frontal, parietal, temporal, and occipital lobes (Figure 2.8). The central sulcus, the demarcating line between the frontal and parietal lobe, is important as it is the source of control of skeletal muscles on the opposite side of the body and the receptor of sensory information from the contralateral half of the body. The parietal lobe is responsible for skin, muscle, and joint sensations, while the occipital lobe is responsible for vision. Hearing is primarily a function of the temporal lobe.

The sense organs transform the diverse sensory information into discrete electrical signals that various nerves carry to the brain. On the way, relay stations filter, sharpen, and modify the signal. The brain is responsible for the formidable task of decoding the electrical signals and extracting meaningful information from it. In addition, information from different sensory modalities has to be integrated and processed at a higher level. According to the famous

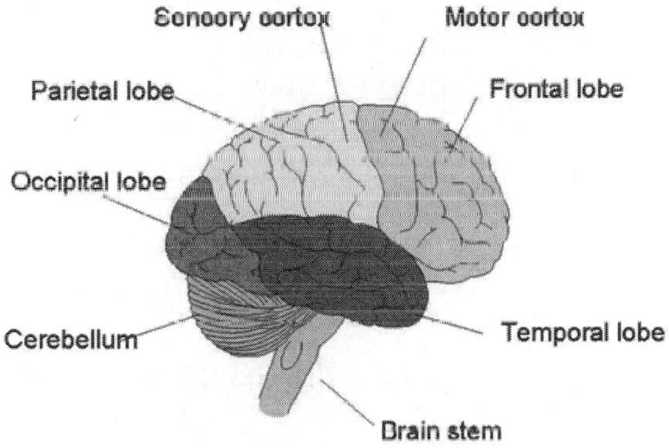

FIGURE 2.8 *Cerebral cortex, showing various lobes on the lateral surface.*

neurologist D. F. Benson, sensory processing can be subdivided into four reasonably distinct steps: perception, discrimination, unimodal association, and heteromodal association.

The peripheral sensory organs carry out perception. At various relay stations between the sense organ and the brain, some data processing does occur. Discrimination is the identification of discrete sensory stimuli within each modality. This includes pain, heat, cold, and so forth from the skin; tone, pitch, rhythm, and the like from the ear; color, size, shape, and other such information from the eye, and so on. The next step of unimodal association involves matching the processed sensory information with relevant information from the memory stores. This leads toward recognition within a single modality. Recognizing a face from a picture and identifying a song while listening to the radio are examples. The last heteromodal modal association is the pooling of information from different sensory modalities and memory stores, and deriving a final conclusion from the total information received. Recognizing a rose by its feel, appearance, and characteristic fragrance will involve heteromodal association. When examining a rose, the skin will detect softness in its petals, while the nose will identify its fragrance. The eyes will register its color, size, and shape. The sense organs will convert the information they have gathered into electrical signals, and the corresponding nerves will carry it to the brain. The brain will process the information and combine the report of the different senses. Finally, it will compare the final image with memory and come up with the best possible fit. This will often necessitate some filling in, stretching, and distorting.

Variation in Perception

As is evident, the whole process is delicate, complex, and cumbersome. Information can be lost, distorted, and misinterpreted during different stages of perception. Contrary to popular belief, perception is highly subjective. The brain attempts to make sense out of the crude sensory information decoded from the electrical impulses it receives. Expectancy, anticipation, and mood, and so on determine perception to a considerable degree. What the brain sees and what falls on the retina are often not identical. There is a whole lot more to seeing than simple image reconstruction.

A number of factors, some of which are known, can distort perception. Hallucination is perception without object and illusion; in other words, it is sensory misinterpretation—the misidentification of a rope as a snake, in dim light.

While a great deal is known about normal perception, much less is known about the mechanism underlying hallucinations. Hallucination in a normal person is by no means a rarity. Neurosurgeon Wilder Penfield from the Montreal Neurological Institute applied electrical stimulation to the brains of awake patients during temporal lobe surgery. His patients reported solo performances of music or orchestras. They heard voices of friends and strangers. Some also saw the people who were talking to them and the man playing the piano music they heard. Several, but not all other investigators reported similar findings. Adams and Rufkin of the University of California at San Francisco found hallucinations following temporal lobe stimulation, but surprisingly enough, repetition of the stimulus at the same site did not produce the same hallucination. Often, then, the hallucinatory experience was quite different. The association between hallucinations and the temporal lobe is further supported by the well-documented occurrences of hallucination as part of seizure activity originating from the temporal lobes.

There are a number of concerns about the veracity of "normal perception." The stretching, massaging, and filling-in associated with sensory data processing is widely recognized and acknowledged by neuroscientists. We do not perceive everything that is out there, and we perceive things that do not exist. Both can and do happen under normal conditions to normal people.

Sight is the most highly advanced perceptual modality in humans. Vision is simply perception of light. Visible light, however, is limited to a narrow band (ranging from wavelengths around 300 nanometers in the near-ultraviolet to 700 nanometers in the red) within a much wider electromagnetic spectrum. The electromagnetic spectrum covers a wide expanse of wavelengths from 10^{-13} (gamma rays), to 10^5 (long wave) (Figure 2.9). Obviously, we cannot detect most of the spectrum with naked eyes, but now we know it exists. The electromagnetic spectrum also includes AM and FM radio signals and television signals. We can receive these signals virtually anywhere, but only with a radio and a television. The same is true of hearing as well. Humans can hear only a narrow band of sound frequencies from about 20 to 20,000 hertz, with higher and lower frequencies being inaudible.

Although all humans have similar perceptual apparatus, there is reason to believe that the world as perceived is not identical for all. This is admittedly difficult to verify, as tapping into another person's perceptual experience is not easy. In many cases, we have to depend upon the unreliable verbal reports of subjective experiences. However, we do have the ability to test this possibility with certain perceptual functions, such as color perception. It is very difficult, for example, to explain what the color green is except by comparing it to various familiar green objects. It is possible that perception of the green coloration of all green objects for one subject may still be qualitatively different from that of another. Four percent of men and 0.4 percent of women suffer from defective color vision. Many with defective color vision who were diagnosed

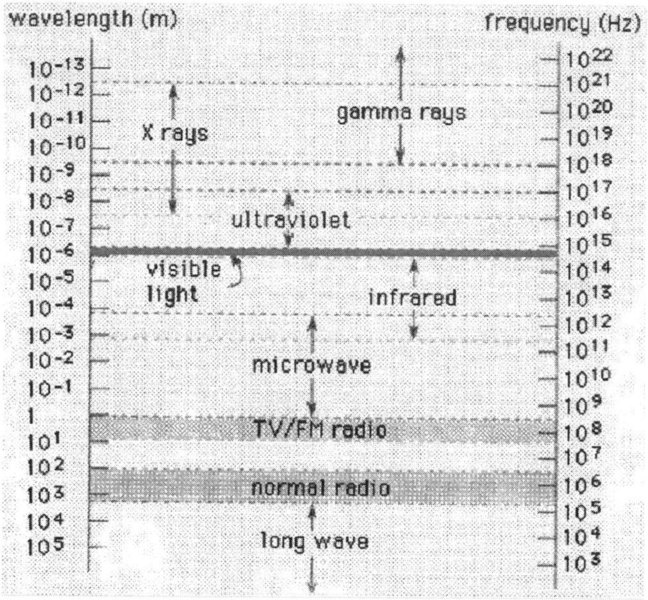

FIGURE 2.9 *The electromagnetic spectrum and the visible light.*

with objective tests did not know that their color perception was different from others.

We do know for a fact that significant intra-individual differences exist in other senses as well. For example, females have a significantly more sensitive sense of smell than males. In 1986, a massive worldwide study conducted by *National Geographic* magazine showed that women in general, especially those from the United States, had high smell sensitivity. European males received the lowest ratings. Substantial differences were found between individuals on what they considered pleasant and unpleasant. Olfactory ability declines with age. A ninety-year-old will be able to identify correctly only one-half of the odors a twenty-year-old can. Similarly, individuals differ in their ability to hear mild nuances in music and their ability to appreciate subtle taste. Younger individuals can hear frequencies the elderly cannot hear. Worded differently, the world as perceived by different humans is likely to be different. We erroneously assume, however, that the world is similar to all.

We share our world with a multitude of life forms that also rely on receiving adequate information about the environment for existence. Their perceptual abilities differ significantly from ours. While some animals can barely distinguish light from darkness, others are able to see objects in very great detail. For example, an eagle perched on the top of a mountain will be able to detect a rat or a snake far below with a great deal of accuracy. Bees and birds can see polarized light and can orient by it. Pollinating insects discern the color of flowers very differently than do humans. Snakes can see infrared, whereas humans cannot.

Bats and whales use sound for navigational purposes. Storks are believed to navigate by use of electromagnetic lines. Snakes use their taste sensation to find food sources and water. A pig has approximately 5,500 taste buds, a cow 85,000, an antelope 80,000, and humans only 4,000. Dogs can hear sounds of a higher frequency up to 40,000 hertz, bats up to 100,000 hertz, and dolphins up to 130,000 hertz, whereas humans do not hear over 30,000.

The human nose contains around 5 million sensory receptors, whereas dogs have around 200 million. Fish and amphibians use a lateral line system for pressure perception. It is very clear that the world is very different to different species. The anthropocentric view that the real world is what humans perceive is grossly misguided.

With advances in technology, we are able to detect signals we could not previously isolate. The electron microscope allows us to see previously invisible objects, while radio telescopes make visible to us celestial events of which we were previously unaware. Technology has given us the ability to produce and receive signals we were unable to detect in the past. In any part of the world, a radio will be able to detect AM and FM signals, which are well below what humans can perceive in the electromagnetic spectrum. Cosmologists have been telling us about events and objects of enormous significance to us and the cosmos that we cannot perceive. The gravitational field of the collapsing star, which generates the black hole, can be so powerful that neither matter nor light can escape it. A beam of light directed at it would disappear, pulled into it. Black holes remain hypothetical since Einstein's days, but recent observations suggest that the star system Cygnus X–1 at the center of the galaxy may contain a black hole. In 1994, the Hubble telescope showed that galaxy MAT 7 contained black holes. The future of our universe is believed to depend upon the amount of "dark matter" present. This substance has weight like all matter but is invisible. It does not have any taste or smell. We simply do not have the technological sophistication to detect it. In fact, the entire field of particle physics and astrophysics deals with phenomena we simply cannot perceive with naked eyes. The hypothesis that reality is what humans can perceive is totally and absolutely untenable.

Perception is identification and interpretation by the brain of some, but by no means all, signals from the environment. These signals consist of a wide range of events, including the photoelectric spectrum, pressure waves traveling through the air, minute chemical particles in the atmosphere, texture and temperature, and so on and so forth. Living organisms, humans included, can perceive a few of such signals. Therefore, different species, even individuals within any one species, do not see an identical world. A dog's world is very different from that of a human. An amoeba perceives the world differently from an eagle. Even human females and males differ in their perceptual capabilities. If matter is that which is perceived, then matter cannot be the same to different organisms. If reality is matter, it, too, cannot be the same. Seeing is definitely not believing.

Now, we know incontrovertibly that the universe is much more than what humans can perceive. In fact, what humans can perceive is only a very small fragment of what exists. The argument can be made that that which falls within our perceptual range is part of a continuum of reality. Although we can perceive only a part of it, matter can still be the reality.

Can there be a reality beyond, besides, and before the material reality? According to the Buddha: "There is an unborn, unbecome, unmade, and uncompounded; for, if there were not this unborn, unbecome, unmade, and uncompounded, there would be no escape from this born, become, made, and compounded" (Udāna, VIII, 3).

The Unseen Reality

The Stream of Time

Perceptual reality is the present; neither the dead past nor the unborn future can be perceived. If perception is reality, reality is a constantly changing phenomenon akin to the music from a record player. The needle is the perceptual apparatus, the revolutions of the record represent the flow of time, and the music is perceptual reality. In both Buddhism and the Upaniṣads (Śvetāśvatara Upaniṣad, I, 4), a wheel (Figure 3.1) depicts perceptual reality. The part of the wheel that comes into contact with the ground is the perceived present, the ground ahead the future, and that behind is the past.

Our point of contact with the stream of time has an internal and an external component that is perception-dependent. Little is known about the internal sense of time and the factors that regulate it. Different states of mind and moods influence it. For example, a number of drugs—marijuana, peyote, and LSD—distort the internal time sense. Time flies when one is elated, and it drags during depression.

The sense of time based on perception is better understood. At least two factors regulate perception of the present: the intensity of the signal that activates the perceptual apparatus and its duration. For a stimulus to be detected, it has to be above a certain threshold in both intensity and duration. Signals that fall short of this thresh-

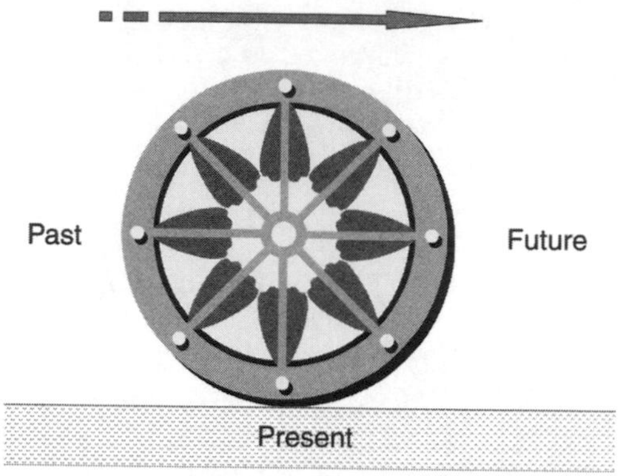

FIGURE 3.1 ***Wheel depicting the passage of time.*** *The part of the wheel in contact with the present is the perceived present, the ground ahead the future, and that behind, the past.*

old will be not detected. This value varies widely from sensory modality to modality, species to species, person to person, and situation to situation. For vision, 2,500 quanta of light absorbed per square centimeter of retina will be perceived by about 60 percent of people. (Quantum is the minimum quantity of light that can be utilized, that is, Planck's constant—6.63×10^{-27} erg-second—multiplied by the frequency of light.) Two stimuli lesser than this value, each unable to cause vision independently when falling on adjacent parts of the retina, can become visible through spatial summation. Similarly, two weak stimuli when presented back to back separated by less than 0.1 sec can trigger vision through temporal summation. In both instances, the two spatially and temporally separate stimuli will be seen as one. In other words, 0.1 sec is the minimum duration of the present, based on visual perception.

The same phenomenon is seen in relation to audition at about 1,000 interruptions/second and tactual vibration at about 4,000 pulses/second. Below this, the sensation will be felt as continuous.

Thus, the perception-dependent sense of time—past, present, and future—varies from sensory modality to modality.

The Relativity of Time, Space, and Matter

Time, in turn, controls perception of space. Perceived objects are located along the three directions of space—vertical, horizontal, and sagittal. Alterations are changes across time, and where there is no time, there is no change. For an object to move from one location to another, the object has to move from one point in time to another as well. Changes in space can occur in any direction, but changes along the time dimension, to the best of our knowledge, are unidirectional—from the past through the present to the future.

Time and the time-dependent space constitute the props and pillars of perception. The substantiality of perceptual reality will therefore be dependent upon the stability of time and space.

Isaac Newton (1642–1727) provided mathematical support for the concepts of absolute time and absolute space. According to him, they served as reference points against which all positions and motions could ultimately be measured. The term absolute is drawn from the Latin word *absolutes,* meaning "isolated." In its present usage, it means that which is primary, independent of anything else, and predates everything else. According to Newton, absolute time was distinct from relative apparent and common time, which were "measured by movements of stars as well as clocks." He saw no connection between space and time. Space was a three-dimensional continuum, and time, a separate single-dimensional one.

These age-old concepts held sway till 1905, when Albert Einstein obliterated this firmly entrenched belief of absolute time and absolute space and the related reality.

Einstein argued that space and time were not inseparable. In fact, he combined time and space into a single "space-time." Neither time nor space, nor space-time, was absolute. All were relative to a single absolute factor, namely, the speed of light. Speed of light is always constant at 186,282 miles/second. Velocity of light was identical regardless of the source and direction. It did not have a reference point, but rather, it was the reference point for everything else.

Time and space are relative to the speed of light, which is constant. If two sources of light, "A" and "B", come on simultaneously, a stationary person located at the center would see them exactly at the same time. However, a person standing closer to light source "A" would see that first and see light from point "B" next. Since light travels at the same constant speed, light from point "A" will reach the individual in a shorter time than light from point "B." The observer's appraisal of the distance of light sources A and B from where he or she is located will be dependent upon the time taken for light to travel from both sources to reach the observer. Thus, both the observer's appraisal of distance from the source and time are dependent upon the constancy of light velocity.

It is said that as a young boy, Einstein was curious about what light would look like if one were to accelerate to its speed. While a person observing a race car from the rope line may see only a blur, the race car driver in the next track accelerating at a comparable speed will be able to see fine details. According to Einstein's theory of relativity, one can never accelerate to the speed of light, for light will always be traveling at 186,282 miles/second in relation to the speed at which the observer is traveling. Thus, it will not be possible to match the speed of light, and therefore, one can never observe the flow of light.

Einstein was able to show mathematically that if an individual were able to approach the speed of light, both space and time would undergo drastic changes. Speed, one must remember, is distance covered in unit time and is therefore time dependent. As the speed picks up, time will begin to slow down. An atomic clock flown around the world in a supersonic jet will register slower time as compared to an atomic clock left behind at the airport. This was actually proven through an in vitro experiment. Although supersonic jets can travel only at about 0.0002 percent of the speed of light, it was possible to demonstrate the time slowing (time dilation). Time dilation is not a phenomenon restricted to time-measurement devices. It influences all phenomena related to time:

the physical, such as sunrise and sunset, and the physiological, such as heart rate, peristalsis, and aging.

An individual on board a vessel traveling at a speed close to the speed of light will age less than his or her twin at home. To the stationary twin, the vessel would seem to shrink in the direction of travel but appear just as tall. As time expands, the length shrinks. To the twin on board the vessel, the changes would seem to happen outside and not within the vessel. As the speed increases, the traveler will see around the corners of passing objects. Things behind will appear within the forward field of vision. As the speed approaches the speed of light, everything will be squeezed into a tiny circular window in front. To onlookers, as the traveler is moving toward them, the traveler will look blue in color (blue shift) and as the traveler bypasses the observer, the color will turn to red. The traveler, as was mentioned above, becomes compressed in the direction of motion. The traveler's mass increases.

An illusion is a sensory misinterpretation. When a rope is misinterpreted as a snake, it is an illusion. Can travel at the speed of light produce illusions of altered time and space, both for the traveler and the observer? Physicists argue in the negative for at least two reasons. When an atomic clock is flown around the world in a supersonic jet and compared with a stationary clock, the former shows the passage of less time. The elapsed time as registered by both clocks is compared at the same time by the same individual using the same perceptual apparatus. One shows less time has passed than the other. The discrepancy between the two clocks cannot be attributed to an error of perception. Particle accelerators can whip small subatomic particles like electrons into very high speeds. The accelerator will have to allow room for the associated increase in mass of the particles. The change is real and not illusory. Secondly, these changes in time as well as space, which are both related to speed, can be demonstrated mathematically. These changes in time and space are predicted by mathematical calculations.

As we have seen, perception definitely depends upon time and space. Nerve conduction, which carries sensory signals to the brain and mediates its processing, is time dependent. Time and space, which are relative, also bind the object of perception and the sensory signals by which it is perceived.

Audition depends upon pressure waves that travel through the air and microscopic particles that stimulate the olfactory sensors. Both are all time dependent and, therefore, relative.

Vision, however, is dependent upon light perception; speed of light, according to Einstein, is the absolute. Thus, if there is reality in perception, it is most likely to be the case with the direct perception of a light source. However, there are problems even here. Matter that produces light (a flashlight) or that light reveals and the train of events that constitute physiology of vision, from the light receptors in the retina to the brain, are relative.

This is probably the reason Einstein called perceptual reality "delusional perception" instead of illusion or hallucination. What is perceived is accurate within the limits of the perceptual apparatus, but the notion that what is perceived is real is a false belief, held in spite of firm evidence to the contrary, and, therefore, delusional.

Although Einstein's special theory of relativity initially sent a shock wave through the scientific community, it was eventually well accepted. However, it suffered from a number of drawbacks, notably the omission of the phenomenon of gravity. Newton's theory of gravity, which held sway before Einstein, seemed to contravene the concepts of the special theory of relativity. Einstein's general theory of relativity, which was put forward in 1915, provided a new explanation for gravitation. The new theory also combined space and time with matter and energy. Space-time was conceptualized as a large sheet made of an elastic material. Planets, globular masses of matter, sank into the sheet and produced distortion of space-time around it.

Gravity was the result of this distortion in space-time. An example will illustrate the point well. A whirlpool can be created around a bowl by immersing it in a body of water. The bowl will be at the vortex of the whirlpool. When another small bowl is floated in the whirlpool, the swirling current will carry it round and round in a trajectory and eventually pull it into the larger vessel in the center. According to the general theory of relativity, a similar mechanism causes the smaller bowl, Earth, to be swept into the swirling eddies caused in space-time by the larger vessel, Sun, around it. In other words, gravity is a consequence of the warp or curve in space-time caused by uneven distribution of energy (matter) in it.

The bend in space-time also dictates planetary revolutions. Einstein's general theory of relativity was, by and large, compatible with Newton's theory of gravity, but it explained the phenomenon in a different way. There were instances where Einstein's theory and Newton's theory disagreed. Experiments showed that Einstein's theory was more accurate. According to Einstein, anything that crossed the dimple in space-time, including a beam of light, would bend. According to this, light from the stars beyond the sun would be bent by the vortex in time-space caused by the sun. Thus, estimation of the star's position based on the view from the earth will be faulty. This was actually verified in Brazil and Africa during a total eclipse on May 29, 1919.

Another prediction of general relativity is that time is affected by gravity. An atomic clock kept at sea level and thus more subject to the gravitational force runs slower than that kept on a mountaintop. Of a pair of twins, the one that lives on top of a mountain where time goes faster will age more than the one who stays on a seagoing yacht. If one of the twins chooses to go on a starship traveling near the speed of light, that twin will be significantly younger upon return in comparison to the one who remained on earth. This is known as the twin's paradox.

Newton could not conceive any relationship between time and space. Absolute time according to Newton was the same all over the universe. Time rolled on at the same pace all over the cosmos. Einstein's main finding was that speed of movement would affect time. A clock when it is moving fast in space will slow down. Speed of travel would also affect mass. The mass of the car will increase as its speed increases. This increase in mass is simply due to the transformation of energy into mass. As an object approaches the speed of light, its mass rises rapidly. An object traveling at approximately 10 percent of the speed of light will increase its mass by 0.5 percent; at 90 percent of the speed of light, its mass would double.

Einstein expressed this relationship between energy and mass in his famous equation, $E = MC^2$, where E is energy, M mass, and C^2 speed of light squared. Since the speed of light is an enormous number, 186,282 miles/second, a small amount of matter is equivalent to an enormous quantity of energy, which finds expression in the explosion of a nuclear device. An object will lose its mass by emitting energy in the form of light.

Concepts of time and space can be confusing, especially to a lay person. In this era of television shows and movies that deal with space travel, one wonders about the outer limits of space "where no man has gone before." Since both space and the spaceship are tangible realities, a question concerning the outer limits of space would also seem to be a very reasonable one. Yet we are not able to conceptualize a boundary on the other side of which there is absolutely nothing. To the human mind, even nothing has to be something. We encounter the same problem with the concept of time as well. Will time continue on forever? It has to since we cannot conceptualize the end of time, because time is the basis for reality, as we understand it. Since it is difficult to comprehend the outer bounds of time and space, we may need to go in the opposite direction to track them to their origins.

According to the general theory of relativity, time, space, and matter are tightly intertwined and interspersed, and therefore, the three should have identical origins. The big bang model, the widely held theory of evolution of the universe, is totally compatible with Einstein's general theory of relativity. This theory, which was originally proposed in the 1920s, was developed into its modern version in the 1940s. According to this model, the universe exploded from a highly dense concentrate of all matter that presently exists in the form of stars and galaxies. The event also represented the formation of time and space. To question the origin of the highly dense concentrate of all matter is nonsensical since time did not exist before the big bang. According to most scientists, this incident occurred 15 billion years ago.

Stephen Hawking and Roger Penrose, in 1965, showed that the universe started with a "singularity," a mathematical concept. In this case, singularity is the point where time and space begin from a total void, in the total and full sense of the expression. At a singularity, space and time and every phenomenon related to it, including physics and mathematics, were nonexistent. In the very early stages, seconds after its origin, the universe was of intense density and extremely hot. From the intensely hot pinpoint, the universe expanded like a bubble to reach its present proportions with all the stars, galaxies, and black holes.

The expanding balloon of time and space contains in its bowels certain vacuous holes, or black holes. The American physicist John Wheeler coined the term "black hole" in 1967. Within these islands of nothingness, time and space cease to exist. The gravitational pull exerted by the black hole is so intense that even light cannot escape it. With no light coming out, these black holes cannot be seen. Their existence can only be inferred by their gravitational effects on the adjacent stars. The black hole is a singularity similar to the big bang. Our own galaxy is believed to contain over 100,000 million black holes. The death of a star, 500 times bigger than our sun, is estimated to culminate in a yawning black hole. The star collapses upon itself, and the intense gravitational forces exerted upon itself creates the extreme state called a black hole.

Einstein's equations related to the general theory of relativity portended an expanding universe, which troubled him. He was more comfortable with a static world of stars and planets and was uneasy with the idea of an unstable expanding universe. In fact, he introduced a "fudge factor" into his equations to eliminate the ballooning of the cosmos. The physical proof for an expanding universe came with the development of high-powered telescopes. In 1908, a 60-inch reflector was built at Mt. Wilson and in 1970, a 100-inch telescope was constructed at the same site. Edwin Hubble, who came to work at Mt. Wilson in 1927, reported that the galaxies were speeding away from each other at a speed proportional to their distance from the Milky Way. Einstein was pleasantly surprised to learn that his equations were more accurate than he himself had thought. (He called his lack of faith in his work the worst mistake in his career as a scientist.) This firm finding of an expanding universe adds further credibility to the notion of an infinitely small beginning, but it also raises questions about the future.

Several hypotheses have been put forward, although no final and firm answers are yet available. Some argue that eventually the expansion will be replaced by a contraction, when the flow of time will be reversed. The universe will go back to its very beginning and disappear through a "big crunch." Other theories favor a universe that continues to expand and cool into the "big chill." Yet others believe that it will expand to a certain point from which it will maintain status quo. Interested readers should consult any of the excellent books written by physicists more familiar than I am with the topic. It needs to be pointed out and emphasized that since time, space, and matter are predicates of an unknown absolute, both creation and destruction of the cosmos will also have to be relative. The absolute unsullied by time and space will have to be immutable, beyond both creation and destruction.

Regardless of the fate that awaits the cosmos, it would seem reasonably clear that time, space, and matter are closely intertwined and that all three are relative. Time and space and their predicates are restricted and limited to the big bubble, which contains the cosmos, as we know it. All sciences exist only within the bubble. If time, space, and matter are relative, everything that is dependent upon time, space, and matter, including Earth, its inhabitants, humans, and the human brain, is also relative. Physical factors such as speed, which can influence both time and space, can also influence the human brain and phenomena related to it. Since time is subject to a number of forces including gravity, the flow of time is likely to be different in different planets across the cosmos. This would indicate that brain mechanisms would also proceed at different rates in different parts of the cosmos. All manifestations of neurophysiology, including perception, will, in all likelihood, be altered. Perceptual reality will undergo drastic changes from planet to planet depending upon the prevalent standards of space-time. A reality that fluctuates from spot to spot can hardly be reality.

It would seem that modern physics after Einstein provides minimal, if any, support for basic tenets of materialism. Since matter is interchangeable with energy, the argument that matter is the ultimate reality stands on weak ground. Perception (of matter), dependent on time and space, cannot be relied upon. The perceived world around us, therefore, cannot be real. Life and death that are dependent upon neurophysiology based on time, space, and matter, cannot really be the true beginning and end. According to the Yin-Yang theory, the world is made of opposites. Everything is relative and made up of polarities with few, if any, absolutes.

Spirituality: Indian Perspectives

The absence of scientific support for materialism does not automatically mean than its antipode, spirituality, is supported. One must keep in mind the spiritual tenet that the absolute reality is beyond and before science and, therefore, science cannot prove or disprove it. Science is a small component of a larger reality, and science validating reality is like a fish proving the existence of the ocean. Those who require hard proof will be disappointed. However it is possible to make some headway by analyzing common experiences with help of reason and logic. An open mind is essential.

The discipline of spirituality embraces a motley of highly diversified schools of philosophy, Indian thought being only one of the many. Indians certainly do not have a monopoly on spirituality. I am an Indian and am thus best acquainted with Indian philosophy, and these are the only reasons for considering it here. I do not believe I will be far wrong in stating, however, that Indian thought antedates most other schools. Although some of the thoughts and ideas seen in Indian philosophical works were also put forward by contemporaneous Greek thinkers, they were not explored and expanded over several hundreds of years in the same depth and detail as in India.

The Vedas, dating back to 1500 B.C. represent the earliest records of human effort to decipher the mysteries of existing and existence. Here one sees the beginnings of the eternal struggle between blind faith and unyielding curiosity, reckless excursions and fearful withdrawal, intuitive experience and logical reason. The awe and wonder of existence alternates with the need and desire to understand and to explain.

Although the Vedas are predominantly theistic, in some verses, intellectual curiosity breaks through. Some lyrical hymns extol Indra, the king of gods, while other parts of Ṛg Veda question his very existence: "There is no Indra, who has ever witnessed him? For whom are these songs of praise?" (8, 100) Ṛg Veda does not describe an anthropomorphic superhuman being who created the cosmos with a single divine fiat. It refers to a single all encompassing Absolute—"that one," Tad-Ekam, devoid of qualities and attributes, without external or internal limits. The first indication of the perennial source of gnawing uncertainty between subjectivity and objectivity can also be found here.

Ṛg Veda starts with consciousness, which it argues antedated the manifested world. This confusing approach begs clarification. Solipsism has a strong presence in Indian thought. Without the subject there can be no object, but subject can exist in the absence of object. Thus, subject will have to be created before object. Consciousness enlivens the subject and therefore has to have primacy over everything else.

According to Ṛg Veda, consciousness was wrought in the fire of Tapas: "*Tapas* transformed Truth into consciousness." The word "*tapas*" literally means "heat." However, it is usually meant to indicate not physical heat but the effulgence of pure consciousness accessed through such austere practices as yoga and meditation.

The first consciousness that was formed, presumably, was that of the creator. The creator, thus, was the product of creation and not the other way around. Before creation, according to Ṛg Veda, there was: "Neither nonexistence nor existence. There was neither the realm of space nor the sky, which is beyond. There was neither death nor immortality. There was no distinguishing sign of night or of day. Darkness was hidden by darkness in the beginning, with no distinguishing signs" (10, 129).

The creator's consciousness pervades the entire universe. Indians (and many other cultures such as the Alaskan Eskimos and Native

Americans) believe that the entire cosmos is self-aware. This concept of a universal consciousness of which human consciousness is part, permeated all aspects of Indian life, including even jurisprudence. Manu the ancient law giver (the Indian equivalent of Moses) declared: "The sky, earth, waters, moon, sun, fire, Yama [devil], wind, night, dawn, dusk, and justice all witness the carnal's actions" (Manusmṛti, 8, 84, 200 B.C.). The subsequent unfolding of the process of creation was a conscious, motivated process and not mechanical and discursive, as materialists would have it.

Once the mind of the creator was formed, it generated *kāma,* or "desire." The word "desire" should not be confused with human desire for acquisition; it reflects the impulse to create. The following steps of creation as the Ṛg Vedic hymns give them are less profound and more pedestrian. The Vedic sage readily admits that all this is speculative and that exact knowledge and understanding of the process of creation is beyond human capability. This disclaimer held sway well into twentieth-century cosmology.

Ṛg Veda refers to these cosmic principles as the *ṛta,* or rhythm, when translated into English. *Ṛta* is closely related to the Primary Principle and is, therefore, immutable and transcendental and eternal. *Ṛta* is present in the weak and strong nuclear forces, electromagnetism and gravitational force, and the presumed grand unified equation, which underlies all manifested phenomenon. It can be seen in the twinkling of a star and the glitter of a diamond. It can be heard in the roar of a thunderstorm and the song of a canary. It can be smelled in the heat of a boiling volcano and the fragrance of a rose. All aspects of creation, both progression and dissolution, are ruled and regulated by *ṛta* and not chance.

The cosmos is not a churning cauldron of random events. It is more than a blind fury of purposeless activity, it is ordered and harmonious around the divine logos. Distant echoes of the Ṛg Vedic verses can be heard in the voice of twentieth-century science. According to Einstein, "Everyone who is seriously involved in the pursuit of science becomes convinced that a spirit is manifest in the laws of the Universe—a spirit that is vastly superior to that of man." Stephen Hawking offers the following view: "The law of science as we know then contains certain fundamental numbers, like the size of the electric charge of the electron and the ratio of the masses of the proton and the neutron. We cannot, at the moment, at least, predict these numbers from theory—we find them by observation. The remarkable fact is that the value of these numbers seem to have been very finely adjusted to make possible the development of life."[1]

The Creating Principle expresses its presence through the principles of creation. Physics, mathematics, and art, ethics, righteousness, and compassion are not divine creations; they are the divine. The creator can be seen and understood through the principles of creation. The scientist must go beyond the frozen, lifeless equations to find the vibrant life and buoyant intelligence that underlies it. According to Ṛg Veda (8, 58, 2), "Penetrate deeper to know the truth, learn first the external and then the internal. He who knows the superficial bond of things with shape, color, and words knows only the outer form of the universe and does not know much. He who delves deeper and perceives the bond within the bond, the tendrils binding separate forces into a larger whole, knows the Real."

The concept of an Absolute Principle, uncreated and unmanifested, yet omnipresent, is developed further in the Upaniṣads, which came later. God is not separate and independent from the world; the divine element is present in every creature and creation, and in every law and principle upon which the created world is based.

The Upaniṣads called the nameless and formless Absolute, "Brahman," meaning that which grows or expands. Several Upaniṣadic passages are given in the form of dialogues. Taittrīya Upaniṣad (III, 1, 1), composed some time after the fifth century B.C., gives a father's response to his son's inquiries of the nature of Brahman: "That from which things are born, that in which when born they live, and that into which they enter at their death is Brahman." Brahman is both the material and mechanic cause of the cosmos. It is the compendium of the creator, creation, and creating. It is the Absolute. Nothing predates it; nothing is external to it, and nothing is primary to it. It comes before everything; everything comes from it. It is both anterior to space and occupies everything completely within and without. The all-encompassing, primary nature of Brahman is summarized in Bṛhadāranyaka Upaniṣad, which was composed before 500 B.C. "That is full; this is full. From fullness, fullness proceeds. If fullness is taken away from fullness, fullness remains" (V, 1, 1). ("Full" was translated from the Sanskrit word *pūrṇam*. There is no exact English equivalent for the word; it can also be translated as "flawless" and "complete.")

Brahman is omnipresent; it lacks in boundaries, dimensions, and divisions. According to Śvetāśvatara Upaniṣad, "It is seen as beyond the three kinds of time, past, present and future, and thus without parts (space)" (VI, 5).

Can this omnipresent ethereal entity be witnessed? Obviously, the transcendental vastness of the Primary Principle will not lend itself to simple human perception. Śvetāśvatara Upaniṣad explains: "His form does not exist within the range of vision, nobody sees this one with the eye" (IV, 20). Bṛhadāranyaka Upaniṣad adds: "It is neither gross nor fine, neither short nor long, neither glowing red [like fire] nor adhesive (like water)." It is not with "night shadow nor darkness, neither air nor space, unattached, without taste, without smell, without eyes, without ears, without voice, without mind, without radiance, without breath, without a mouth, without measure, having no within and no without" (III, 8, 8).

Yet every single living person is aware of the underlying truth at some level, in all its manifestations. Kena, one of the later Upaniṣads composed after A.D. 200, expands this concept in several passages. "It is what cannot be spoken but can be spoken, what cannot be thought, but can be thought. What cannot be seen but can be seen, what cannot be heard but can be heard, what cannot be smelled but can be smelled, know that alone to be the spirit. The Absolute is implicit in everything but not completely in any. Those who say that they do not know it do know it, and those who say they know it for sure, do not" (I, 2–9).

To many, the concept is too diffuse to be useful. Chāndogya Upaniṣad, provides a simple explanation during a conversation between Uddālaka Aruni and his son, Śvetaketu. Śvetaketu was curious as to how "the unhearable becomes heard, the unperceivable becomes perceived, the unknowable becomes known?" Toward the end of a lengthy explanation, the father asked his son to put some salt in a pot of water and to return the next morning. The next day when they were together, Śvetaketu could no longer see the salt. "Please take a sip of it from this end," the father instructed. Śvetaketu did so and could taste salt. "Now take a sip from the middle. How is it?" the father asked. "Salt," said Śvetaketu. "Again, take yet another sip from the other end. How is it?" "Salt still," Śvetaketu repeated. "Stop, throw it away and come to me," the father said. "You did not see the salt but it was there everywhere. Similarly you do not perceive the Pure Being here but it is everywhere" (VI, 13, 1–3).

One does not need to traverse the whole world to find Brahman. One need not go to India or Tibet, and one need not become a Hindu or Buddhist. Elsewhere in the same Upaniṣad, additional explanations are given: "By one lump of clay all that is made of clay becomes known"; and "By one nugget of gold, all that is made of gold becomes known." The Absolute is present in everything that is relative and by knowing any one thing in depth you get to know the Absolute. Both "clay" and "gold" are simply words; in order to know the Absolute, one has to go past the external qualities (color, shape, texture, luster, and so on, which their names represent) and be in touch with the basic substance, which is the same for all (VI, 2, 4–5).

The point here is that all objects are made of the same substrate, and by understanding any one, you get to understand the substrate for all. The principles of physics are the same in Mars and Earth. At a deeper level, energy, which constitutes matter, is one and the same, all over the cosmos. According to Chāndogya Upaniṣad:

> *Just as, my dear, bees prepare honey by collecting the essences of different trees and reducing them into one essence. And, as these "juices" possess no discrimination (so that they might say), "I am the honey of this tree, I am the honey of that tree," all creatures though they are composed of the same essence, do not realize that they are. They see themselves as whatever they are in this world, tiger, lion, wolf, boar, fly, or gnat, worm, or mosquito. That*

which is the subtle essence is the same for this whole world. That is the truth. That is the self. That are thou. (VI, 9, 1–4)

All human beings are made of the same material of which the cosmos is made. The same principle that applies to matter applies to humans as well. This principle also applies to the brain, which generates awareness of self. By understanding that which constitutes self, one is able to understand the entire cosmos. Thus, the Upaniṣads argue that the key to understanding the cosmos lies within one's self. The phrase "that art thou" *(tat tvam asi)* (Chāndogya Upanisad, VI, 8, 7) is regarded as one of the most profound and meaningful passages in Indian philosophy (Mahāvākyā).

Of two individuals gazing at the night sky, one may be enthralled by the loveliness of the star-studded firmament, while the other may just see shiny white dots against a dark background. The images that fell on both individuals' retinae were identical. However, processed sensory information, the conclusions arrived at, and, most important, the emotional responses to it were very different. I do not know how we can enable the person who saw nothing more than bright holes in a black pall to see and experience the profound majesty and magnificence of the star-studded night sky. Similarly, I also do not know how to respond to an individual who cannot sense the creator in the creations. Radhakrishnan writes, "It is the same supreme reality which lives in all things and moves them all, the real one that blushes in the rose, breaks into beauty in the clouds, shows its strength in the storms and sets the stars in the sky."[2]

Advaitam: Explorations of Reality

The branch of Indian philosophy that integrates and synthesizes these thoughts and ideas into a composite whole is called Advaitam (or Advaita). Dvaitam means dual and Advaitam, non-dual. The basis of this school of thought is the concept that the basic substrate for the entire cosmos is one and the same. The Indian philosopher Vivekananda (1863–1902) explains: "The Absolute has become the universe by coming through time, space and causation. This is the central idea of Advaita."[3]

Advaita Vedānta literature goes back at least to A.D. 550, and over the years, a number of authors contributed to it. Although proponents of the school claim uninterrupted, seamless continuity with the Upaniṣads, most experts identify Gauḍāpada as the first systematic exponent of the system. He came up with the vast majority of the key concepts, which the later authors expanded and elaborated. Gauḍāpada credits the Māṇḍūkya Upaniṣad (composed around 500 B.C.) as his major source of inspiration and also mentions the Buddha by name in his celebrated book, Gauḍāpadiyakārikā (also called Madukyakārikā). The antagonism between Buddhism and Hinduism, which developed in later years, was not prevalent at the time of Gauḍāpada. Not much is known about the life and times of Gauḍāpada except that he lived in present-day Bengal (also known as Gauḍādeśa) in the northeastern part of India about A.D. 550.

Gauḍāpada stated in clear and rational terms that the external world conditioned in time, space, and cause cannot be real. He rejected out of hand the materialistic view that matter is real. He questioned the reliability of our perception and that of the others. He believed that memories and thoughts based upon objects of perception would have to be considered unreal for the same reason. He dismissed language, which describes the perceived world, as unreal.

Dreams, which appear real while dreaming, are dismissed as unreal upon waking up, based on the grounds that the dream had a beginning when one went to sleep and ended when one woke up. In other words, it is dismissed as unreal because the content of the dream phase lacked continuity with the waking state. It is suspended in thin air with nothing preceding or succeeding it. Waking life is accepted as real, although it also has a beginning at the time of birth and ends at the time of death. We do not know where we were before birth or where we will be after death. It is discontinuous with what was there before birth and probably what will be there after death. For all we know, it too hangs loose in an unknown vacuum. Yet it is accepted as real.

Having established the nonsubstantiality of time, Gauḍāpada argued that time-bound changes also have to be nonsubstantial. Change is alteration over time, and if time itself is unreal, how can change be real? Gauḍāpada also rejected the notion that the absolute universal spirit produced the world of manifestations on the grounds that production involved change. Both creation and destruction are forms of time-dependent change and, therefore, cannot be real.

What is created is on closer examination found to be nothing but an illusion. In truth, no change has happened or can happen. There can be no discernible purpose in the creation of the illusion except that it is the inherent nature of the creating agency. According to the Yin-Yang principle, creation implies destruction, that is, a start and an end. If one accepts creation and destruction, one should also accept the argument that there was nothing before creation and there will be nothing after destruction. If there was nothing in the beginning and there will be nothing in the end, in all probability there is nothing in the middle, either.

Thoughts and ideas put forward by Gauḍāpada were clarified, expanded, and consolidated by Śāṁkara (A.D. 788–820), who is undeniably the life and soul of Advaitam as it is understood today (Figure 3.2). Śāṁkara was a prolific writer, and the exact number of his literary works is a matter of controversy. In his hands, nascent Advaitam grew and burst forth into a well-rounded system of philosophy that eclipsed most of the other schools. Śāṁkara and Advaitam were largely responsible for the demise and disappearance of Buddhism from India. Ironically, the views Gauḍāpada and Śāṁkara espoused were very close to Buddhism, especially the Madhyāmaka and the later, Vignānavāda schools. Śāṁkara's system of philosophy was remarkable for its comprehensiveness, logical subtlety, and intellectual sophistication. It bore witness to his unusual ability to combine intuitive insights with logical reasoning and support one with the other.

The illusory nature of the manifested world is a central theme in Advaitam. The multifarious shapes and forms that characterize the phenomenal world are creations of *māyā*, which cloaks and con-

FIGURE 3.2 Śāṁkara (A.D. 788–820).

ceals the true, monistic nature of reality. The Absolute, according to Śāṁkara, is: "That which permeates all, which nothing transcends and, which like the universal space around us, fills everything completely from within and without."[4] Uncritical acceptance of *māyā* is called *avidya,* or "nescience."

The phenomenal world is simply the distorted, unreal manifestation of a primary undivided reality. Śaṁkara uses many illustrations to clarify the point. An individual with defective vision may see two moons. That does not mean that there are two moons. The perception of duality is due to impairment. After a deluge, the moon may be seen in many puddles, which does not mean there are many moons. There is only one moon, the reflections of which may be seen as many. Foam, froth, and bubbles appear different and are known by different names, although all are constituted by water and air.

If the universe is unreal, then its creation will also have to be unreal. Why should the real masquerade as unreal? Śaṁkara dismissed a number of explanations from Vedic literature as untenable and concluded that the purpose behind the unreal manifestation of the real was beyond human comprehension. Questions concerning when the transformation occurred are meaningless since the concept of time is applicable only to the finite and not the infinite. Similarly, to question whether only a component of the infinite transformed into the finite is also irrelevant, since space does not apply to the latter.

The world is made of opposites, and denial of reality can be made only in reference to something that is real (Brahmasūtrabāṣya, III, 2, 22). The term "illusion" cannot have any meaning except as the antithesis to the real. When physicists argue that time and space are not real, they are also inferring that something else is.

The Absolute is totally devoid of qualities and, therefore, cannot be perceived. Since it is beyond time, it cannot have a beginning, a middle, and an end. It cannot be compared to anything else, since everything has its basis in it. It does not have an ego (self), since ego cannot exist without non-ego and the Absolute cannot be divided. It is neither a subject nor an object; both have their origins in it. The Absolute cannot be thought of or imagined, and that which cannot be thought of, cannot be spoken or written. However, experience antedates thoughts; the Absolute can be experienced.

Reading about the Absolute does not give you the experience. Śaṁkara insists: "Sickness is not cured by saying the word medicine. You must take the medicine. Liberation does not come by merely saying the word Brahman. Brahman must be actually experienced" (Vivekacūḍāmaṇi, verse 62). Studying scriptures, performing rites and rituals, and engaging in endless prayer are also not much help. According to Śāṁkara: "Study of scriptures is meaningless in the absence of the experience of the Absolute. And after the Absolute is experienced, study of scriptures is still meaningless" (Vivekacūḍāmaṇi, verse 59).

In order to be in touch with the real, we have to cut off attachment to the unreal. So long as we are tethered to the unreal, we will not be able to get past it. Śāṁkara makes it very clear that the experience of reality comes easier for some than others. Everybody cannot be proficient in mathematics, but some are more gifted than others. Similarly, all cannot become virtuosos in music or even discriminating aesthetes. That, too, will depend upon talent, which is, by and large, inherited. The same applies to spiritual experience. Śāṁkara does not provide detailed description of those who cannot, except to note that they are "slow-witted" *(mandabuddhinam).*

Śāṁkara argues: "A clear vision of the Reality may be obtained only through our own eyes, when they have been opened by spiritual insight—never though the eyes of another. How can we see the moon though another's eyes?" (Vivekacūḍāmaṇi, verse 54).

Consciousness and Its Veils

The Light of Consciousness

Sensory and extrasensory perception take place in the brain.

Both reality and unreality reside there. Materialism and spirituality and the inability or ability to experience the unseen Absolute depend upon neural mechanisms. Śaṁkara declares, "The mind is the place of high sanctity where scripture, sacraments, and sacrifices unite with the Gods" (Upadeśasāhasrī, XIV, 40).

The term "mind" appears to have more currency in philosophy than in science. Although occasionally one runs into the term in scientific literature, it is seldom, if ever, defined with the precision and objectivity that characterize science. Predominantly subjective, the concept of mind is lacking in borders and boundaries of any sort—spatial, temporal, or otherwise. A quest for the outer boundaries of the mind would be just as futile as that for the outer boundaries of space or the beginning of time.

Comparisons between the mind and matter have vexed the human mind since antiquity. In India, the issue was first addressed by the Sāmkhya system, the oldest school of philosophy in India. Some of the Sāmkhya concepts can be traced all the way back to 1500 B.C. Perhaps beyond. Some scholars believe Sāmkhya had its roots in the Indus Valley civilization.

According to Sāmkhya, the entire cosmos and everything it contains can ultimately be reduced to two basic principles—Prakṛti and

Puruṣa, which roughly correspond to the more modern concepts of matter and spirit. All that is perceived and unperceived, known and unknown, real and unreal, fall under Prakṛti and Puruṣa.

Sāmkhya believes Prakṛti cannot be destroyed, that only transformations are possible. A clay pot of water cannot be expunged out of existence, only altered. Broken fragments of the pot still exist and the water is still present, mixed with the soil. Although Prakṛti evolved and developed into objects that vary in size, shape, color, and complexity, Prakṛti is essentially lifeless and devoid of self-awareness. The human body, including the brain, is derived of matter that, in turn, is Prakṛti. Thus, the brain and its constituent parts (and their behavioral manifestations) are also mediated by this principle. To this extent, Sāmkhya is compatible with contemporary medical and physical sciences. In short, mind is dependent upon the brain, which is made up of matter.

The primordial, unformed Prakṛti transmutes into formed elements through the interactions between its three constituent powers called *guṇas*.

Interestingly, modern physicists tell us that molecules, which derive from atoms, constitute matter and that atoms can be further broken down into protons, neutrons, and electrons. According to the most recent superstring theory, the fundamental constituent of all matter is tiny vibrating strings of energy. Factors responsible for the progression of matter from unstructured energy through structured subatomic particle to matter continue to be a mystery. Sāmkhya proponents of 600 B.C. ran into the same problem.

According to the superstring theory, particles, which make up matter, are different modes of vibration of the string. According to Sāmkhya theorists, Prakṛti forms into three strains, which determine its evolution into more complex structures. These strains themselves cannot be perceived, but their effects belie their presence.

The first of the three *guṇas* is called Satva, derived from the Sanskrit word "*sat,*" which means that which is real, substantial, or perfect. This is the state of equilibrium without positive or negative agitation. Prakṛti inherently seeks out this state.

The second, Rajas, manifests itself as activity.

The third, Tamas, represents the obstacles Rajas strives to overcome, to reach *sat*.

The three *guṇas* apply to all manifestations, both living and nonliving, and enjoy wide acceptance in India, even by individuals who are totally opposed to the other aspects of Sāmkhya philosophy.

Although the brain is wrought of Prakṛti, for the brain to become alive, a presiding power is essential, just as an electric bulb constructed of glass and filament needs electricity to become luminous. The brain, constructed of, and by, inert matter, needs something to blow life into it.

Sāmkhya believes that this life-giving agency is different from, and independent of, Prakṛti. It is Puruṣa.

A candle is created for the specific purpose of supporting a flame. Although the flame is totally dependent on the candle for its existence, the flame is not the same as the candle. The flame cannot exist without the candle. Yet the two are entirely different entities.

Similarly, life cannot exist without the brain and, upon destruction of the brain, life ceases. However, despite life's dependence upon the brain, they are not one and the same.

Puruṣa is definitely not the body, and it is not a product of physical elements. It utilizes the brain for all receptive and executive functions, but it is not the brain. It is that which activates memory and intelligence, but it is not them. The relationship between the brain and Puruṣa is intimate and overlapping, but essentially separate. No function of the brain will have any meaning in the absence of Puruṣa. Puruṣa, on the other hand, will have no material existence in the absence of the brain.

Puruṣa is transcendental and totally nonmaterial. On its own, it does not think, it does not feel, and it does not act. All emotions—happiness and sadness, loving and hating, attraction and aversion—are functions of the brain derived from Prakṛti. Prakṛti also mediates other brain functions, including memory, perception, thinking, and acting. Puruṣa, however, is the agent that receives the input from sense organs, that thinks thoughts, that keeps information in memory stores, and that intimates and sustains all actions.

Puruṣa is in the quintessential subject; it is at the heart of the concept of "self." Like Prakṛti, Puruṣa is also indestructible.

Although Puruṣa is life itself, it can experience life only in collusion with Prakṛti.

Prakṛti is devoid of consciousness, but it is spatial and temporal—evolving, restless, and undergoing changes. Puruṣa is the light of conscious awareness, the ultimate subject, and the ultimate utilizer of the brain and all its capabilities. Yet Puruṣa cannot perceive itself, as the eye cannot see itself. Puruṣa can perceive itself from its reflection in the brain (that is Prakṛti), as the eye can see itself only in the mirror.

Consciousness, which is independent of the brain, can experience its state of being conscious only through the medium of the brain. Through its association with Prakṛti, the otherwise unfeeling, unthinking, nonacting Puruṣa blossoms into a living person, feeling, thinking, and acting.

In that state of intimate bonding with Prakṛti, Puruṣa loses its identity as an independent entity and merges with the brain and its operations. Puruṣa no longer sees itself as the life-giving element of the brain; instead, it believes it *is* the brain.

Bondage, Sāmkhya argues, is the delusion Puruṣa maintains that it is the body and the brain. Release is freedom from this delusion and knowledge of its independence. But this is possible only when the brain, through character modifications and austere practices, is brought into the *sat* stage. When the mind is turbulent with Rajas, or torpid with Tamas, the underlying Puruṣa cannot be separated from the brain and accessed in its unsullied purity, just as the pebbles at the bottom of an agitated muddy pond cannot be visualized. Similarly, the restless mind caught in worldly cares and concerns will be opaque, its depths difficult to visualize. It will reveal the glory within when it is calm, tranquil, and serene. Then Puruṣa shines through with its purity, simplicity, and magnificence.

Liberation, according to Sāmkhya, has nothing whatsoever to do with rites and rituals, penance and prayers, songs and sacraments. It is simply a matter of self-discipline, elevation above the turbulence producing attractions and aversions, and cultivation of practices designed to silence the mind. It is basically and essentially destruction of the idea that you are your body, including the brain.

Sāmkhya holds that the ultimate goal of every living being, whether it is realized or not, is liberation of Puruṣa from Prakṛti.

Since Puruṣa is totally undifferentiated, it is unclear how it can perceive itself at the time of liberation from the brain, which it needs for awareness of any sort. What becomes of both after liberation is left open; admittedly, the issue of "after" may not apply to the timeless Puruṣa.

Sāmkhya vehemently denies the concept of a universal spirit and insists on the separateness of individual Puruṣas. Separation will have to be irrelevant for an entity that is beyond and before spatial limitations. Similarly, it is unclear how the lifeless Prakṛti, through the random interactions between the *guṇas,* can evolve into the highly purposeful entities called bodies—specifically the brain—designed to accommodate and nurture Puruṣas. It is accepted that Puruṣa somehow influences the evolutionary process that leads to the development of the brain, but the details are missing.

Sāmkhya philosophers took pride in being totally rational and logical. They went out of their way to avoid metaphysics and religion. Carried away by their enthusiasm, they attempted to categorize, rationalize, and mechanize all phenomena, even Puruṣa, which Sāmkhya itself accepted as transcendental, and therefore beyond human logic.

Śāṁkara, one of the most outspoken critics, asks how the passive and essentially detached Puruṣa can impel and influence Prakṛti, which is totally independent. Actionless and attributeless, Puruṣa by its sheer proximity cannot goad the inert Prakṛti to undergo the evolutionary changes. Prakṛti is lifeless and the Puruṣa is indifferent, and there is no third factor that yokes them together into an efficient relationship. The overarching problem for the entire system would appear to be the absence of an ontological principle that presides over both Prakṛti and Puruṣa.

The major point is that the brain provides the machinery to support and sustain consciousness, but that consciousness is not a product of the brain, or vice versa. There is no clear way to prove or disprove this hypothesis. Coincidence of consciousness with brain development and its dissolution with brain death do not necessarily mean a dependent status of consciousness with the brain. As was mentioned earlier, when the candle wax has completely melted off and the flame disappears, that does not mean that the flame was a product of the candle. When gasoline runs out, the car stops. That does not mean that the car was a product of gasoline, or vice versa. Life is not possible without nourishment. That does not prove that life is a product of food.

Although Sāmkhya probably predated the arrival of Āryans, the Vedic literature was developed and popularized almost exclusively by Āryans who occupied the upper castes, especially by Brāhmins, the priestly class. The nontheistic components of their Vedic literature show strong Sāmkhyan influence. The Upaniṣadic sages, however, embellished these ideas, filled in the gaps, and supported it with an ontological emphasis.

The *Kośas*

Among the various descriptions they provided, the system mentioned primarily in the Taittrīya Upaniṣad (500 B.C.–A.D. 200) is best known and most widely accepted. According to that Upaniṣhad, *kośas,* or vestures, ensheathe the subjective self, referred to as Ātman (Figure 4.1). Derived from the Sanskrit word "an," meaning "to breathe," Ātman means life or soul. It can also stand for consciousness and the divine spark of Christianity. All Upaniṣads accept Ātman as the fundamental life-giving principle, beyond time, space, and causation. (It should be understood that the ensheathing is functional and not anatomic.) The *kośas* represent the body *(annamaya kośa),* vital functions *(prāṇamaya kośa),* mind *(manomaya kośa),* intelligence *(vijñāna kośa),* and bliss *(ānadamaya kośa).*

Descriptions of these sheathes from 1,500 years ago cannot be translated into modern neurosciences with any degree of exactitude. The terms used and descriptions provided of the constituents are archaic and hard to interpret. In his famed book *Vivekacūḍāmaṇi,*

Śāṁkara, venerated for his lucid expositions of cryptic Vedic texts, explains the *kośas*. I have relied heavily on Śāṁkara's descriptions, but not exclusively. Still, it must be made clear that explanations of their neurophysiological basis are approximate and tangential, at best.

The first *kośa* is *annamaya kośa*, "anna," meaning food. This sheath represents the mental image of the body derived from gross matter. The physical body, which is totally dependent upon food, shrinks and shrivels in the absence of nourishment. Vedic literature distinguishes gross physical body *(sthūla śarīra)* from subtle body or mind. *Annamaya kośa* is not the gross body as such but the image of it, which is part of the *sūkṣma śarīra* that is mind.

The next sheath, *prāṇamaya kośa*, is made of life-sustaining functions of which breathing is principal. Awareness of inner bodily functions includes such events as the beating of the heart and air currents moving in and out, rolling of the stomach, peristaltic contractions of the intestines, a full bladder and rectum, and so on. Most, but not all, of these are "autonomic functions," as modern physiology would have it.

Neurophysiologically, the next two sheathes—*manomaya kośa* and *vijñānamaya kośa*—overlap a great deal. Although *mana* means mind, perception is included here. The Eastern thinkers hold that the mind is an active and definitely not a passive participant in perception. Mind is much more than a mirror in which external objects are reflected. The mental apparatus chooses the object of perception, reaches out to it, encircles it, draws its image into its own self, embellishes it, and presents it to the self.

The Sanskrit word *vijñānamaya* is difficult to translate into English. It can mean intelligence, knowledge, or factual information; however, its source is organs of perception and its contents, empirical. *Vijñānamaya kośa* is made of the knowledge and related mechanisms that include cognition and memory. Indians differentiate intuitive or immediate knowledge *(jñāna)* from discursive or mediate knowledge *(vijñāna)*. *Vijñāna* is information acquired through the senses, processed, and stored. It excludes unassisted, nonvolitional, spontaneous insights. In addition to factual knowledge, it

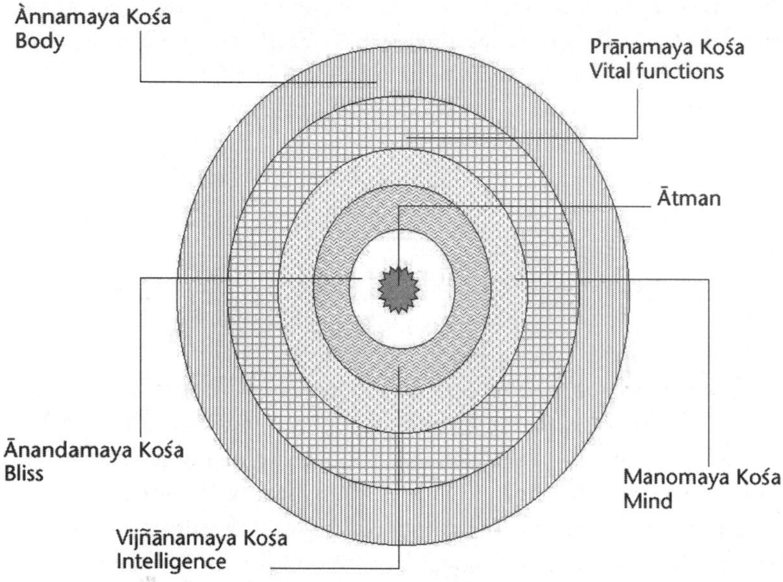

FIGURE 4.1 ***The Upniṣadic concept of the human mind.*** *The mind is constituted of different layers representing the body, vital functions, mind, intelligence, and bliss that surround the central ontological Principle of Ātman.*

includes life's experiences and expectancies. *Vijñānamaya kośa* also includes thought. Setting goals and development of plans to achieve them are also part of this sheath. Emotions, pleasure of success, pain of loss, anxiety of anticipation, fear of failure—all come under this category. Built upon the extremely shaky foundation of the mediated report of the senses, *vijñāna* is weak and unstable.

Physiology of perception, control of vital functions, cognition, volition, memory, thinking, and so forth—functions of the outer four *kośas*—are well covered in the many neurology textbooks. Activities of the *kośas* can be subsumed under two broad categories: (1) the functions, and (2) self, the agent. Under normal conditions, the *kośas* are fused together and perceived as a single, seamless whole, and the agent is usually confused with its functions. For example, many would consider their bodies as part of the agent, or self.

The body image is mainly derived from somasthenic sensations arising from the skin, muscle, joints, and so on. The primary somatic sensory cortex where sensations are perceived is located in the cortical gyrus behind the central sulcus. In the somatic sensory cortex, areas that are more sensitive—tongue, face, hands, and the like—take up more space.

When we cut our nails and hair, they cease to be part of our self. The same is the case with an amputated limb. If amputations were possible at the neck level, I would wager the self would be with the head and neck, provided the brain was kept adequately perfused and alive. Since self is maintained even when parts of the body are removed, the body cannot be the true self. Śaṁkara declares, "The ignorant thinks he is the body" (Vivekacūḍāmaṇi, verse 160). He adds, "Do not identify with the body as you do not identify with your shadow or reflection" (Vivekacūḍāmaṇi, verse 163).

If self is not the body, then what is it? How about the other mental functions? Are they part of self?

There is a subject within every human being who perceives and reacts and who puts to use all faculties of the mind. In addition, the subject may chose to act spontaneously without reacting to any external stimuli. The impetus to action can be exclusively internal, arising from the thought and memory processes. Once the subject focuses on a sensory stimulus or chooses to engage in a certain course of action toward a desired goal, the mechanism of attention is effectuated. We have no clear idea about the neurological basis for self.

Self-awareness has to be based on self-perception of the brain by the brain. If indeed a brain region mediates this function, the activity of that region will have to coalesce the perceiving subject with the perceived object, which in this case will be the subject itself. In other words, the subject and object will have to merge.

No description of self would be complete without considering the term "ego," which Sigmund Freud popularized. In psychoanalytical parlance, ego is the agency that negotiates with the id, the primitive bestial instincts, and compromises with the superego, which represents the conscience and deals with the external world. It represents the functional, operational aspect of self. Ego boundary, which is commonly referred to in contemporary psychiatry, is the line of cleavage between the self and the world. Such a concept will engage all *kośas* thus far discussed, with the exception of the perception of the external world. While the objective world is seldom regarded as a component of ego, very often the body is. Divergent functions of the mind, including consciousness, memory, thought, and perception, are integrated and fused into a single whole, which is perceived as ego. Thus defined, there have to be marked differences between people in their egos. Since individuals do not have identical memories, thoughts, and perceptions, concepts of self that include these functions will show significant differences.

The first four *kośas*—namely, the body image, vital functions, mind, and intelligence, tainted with the manifest world—show inter- and intra-individual differences. However, the fifth *kośa, ānandamaya kośa,* is the same for everybody. *Ānanda* in Sanskrit denotes bliss, which is to be differentiated from pleasure. Mundane pleasure is mediated—related to a perceptual experience or memory. Bliss of the *ānandamaya kośa,* on the other hand, is immediate and is not predicated on sensory perception or its memory. It has nothing to with the empirical world; it is uncaused and totally independent of subject-object distinctions. Unlike pleasure, which can be contrasted with pain, ānanda is unimodal, with no contrasts. It is simply the pleasure of life and living.

David G. Myers of Hope College in Michigan, and Ed Diener of the University of Illinois, collated data from about 1,000 surveys of 1.1 million people on their sense of well-being. According to them, 93 percent of the participants reported feeling good about their lives, and material prosperity had very little to do with their happiness.

According to Indian thought, the cause is in the underlying *ānandamaya kośa.* Vedic seers believe that life in its most basic form is sheer happiness. Shadows are cast upon the inner light of happiness by the overlying *kośas,* which relate to and deal with the empirical world. *Ānandamaya kośa* makes living pleasurable; it may be the explanation for the proverbial love of life every living creature manifests. This happiness is endogenous and totally unrelated to gains, possessions, and successes. It just is. According to Taittrīya Upaniṣad, "For who indeed could live, but for this bliss" (II, 7, 1).

Ānandamaya kośa is the wellspring of all pleasure; it sweetens the other mental operations. The reward that accompanies various activities—making money, falling in love, enjoying music, and so forth—is supplied by the *ānandamaya kośa.* According to Śāṁkara, in dream and wakefulness, it can be glimpsed "during favorable conditions and upon fulfillment of desires" (Vivekacūḍāmaṇi, verse 208).

We do not have to engage in various mental exercises to get a faint taste of the *ānandamaya kośa;* it is possible to reach this ocean of sheer ecstasy and to immerse in it by bypassing the other four *kośas.* This happiness can be experienced in an attenuated form during the happy repose of dreamless sleep (Vivekacuḍāmaṇi, verse 208). When detached from the other four without dulling, as happens during sleep, the happiness of *ānandamaya kośa* is a hundred times as potent as "the pleasure associated with all the gold in the world" (Taittrīya Upaniṣad, 11, 8, 1). Śāṁkara explains, "Who would prefer to stare at a lifeless picture of the moon, when the lovely full moon, high up in the sky fills every heart with unalloyed happiness?" (Vivekacūḍāmaṇi, verse 522). It is this unparalleled pleasure that recluses who spend their entire lives in self-denial are pursuing. "I cannot express by words, or even contemplate, this ambrosia saturated with bliss that I have found. A mere smidgen entrances me and gives me heavenly peace" (Vivekacūḍāmaṇi, verse 482).

The *ānadamaya kośa* is closely aligned with empirical consciousness. Pleasure is fundamental to life even in the most primitive and primordial life-forms. Pleasure and life are intricately intertwined. Although we have no clear scientific support for the pleasure associated with life, we do have some information about the physiology of empirical consciousness.

The Physiology of Consciousness

All tissues are made up of living cells. This includes the muscle, the skin, mucous membranes, bone, cartilage, and so on. Yet these organs, with the exception of the brain, are not self-aware. Life is simply the capacity to engage in functional activities such as metabolism, growth, reproduction, responsiveness, and adaptation to the environment. Life, however, does not mean self-awareness or consciousness, as we understand it.

Consciousness seems to be the exclusive property of the neuron, the smallest functional unit of the brain. Neurons taken from the living human brain can be kept alive in a tissue culture for days and sometimes weeks. They remain alive so long as they are supplied with glucose and oxygen and their metabolic wastes are removed. Hypothetically, an individual's brain could be chopped up into small chunks and each piece kept alive separately, in the laboratory. To the best of my knowledge, nobody has done such an experiment. If the experiment were to be carried out, it would be very difficult to verify whether each piece is endowed with consciousness, which is essentially a subjective experience.

Even if the neuron was indeed conscious, bereft of the faculty of communication, the experience could not be made known. Objectively, in the neuron, life and consciousness cannot be separated. We do not know of an electrical, metabolic, or biochemical index that will distinguish one from the other. It is possible that a critical mass of neurons would have to be assembled in order for the experience of empirical consciousness, as was discussed above, to be generated.

Available evidence suggests that empirical consciousness cannot be produced by a random collection of neurons. Significant parts of the brain may be destroyed or surgically removed without affecting life and the subjective experience of self. The part of the brain that is most indispensable for the preservation of life and self-awareness, as we understand it, is contained in the brain stem. A loosely knit system of neurons called the reticular formation, which extend from the medulla on upward into the thalamus, seems to be essential (Figure 4.2).

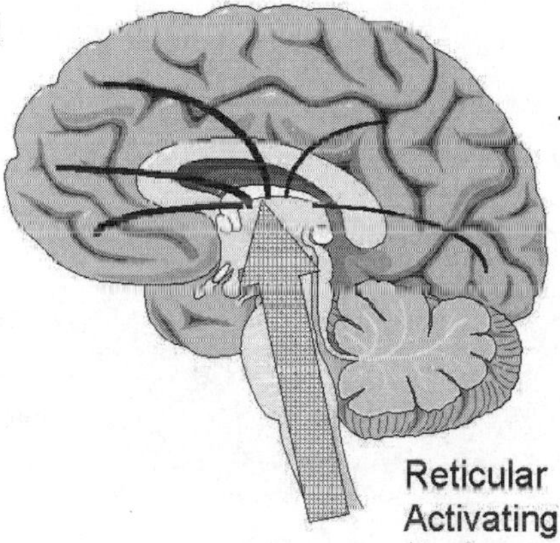

FIGURE 4.2 *A loosely knit system of neurons called the reticular formation, which extends from the medulla on upward to the thalamus and from there to the frontal lobe, mediates arousal.*

Lesions to these brain regions are usually associated with coma, where feeling of self is minimal, if present at all. In animals, lesions on both sides produce a state similar to sleep, with diminished or absent responsivity to sensory stimulation. When one side is damaged, the animals do not respond to stimuli on the other side and seem to become oblivious to what happened on the damaged side. The reticular formation carries out a variety of regulatory activities, including the vital ones of cardiovascular and respiratory regulation. It is also related to sleep and wakefulness and various sensory motor activities.

As they ascend in the spinal cord, the sensory pathways give collateral branches to the reticular formation. Sensory stimuli such as bright light, noises, and pain ward off drowsiness and keep one alert. The connections between the reticular formation that mediates arousal and the ascending sensory pathways may be the explanation. The reticular system has been reported to be most inti-

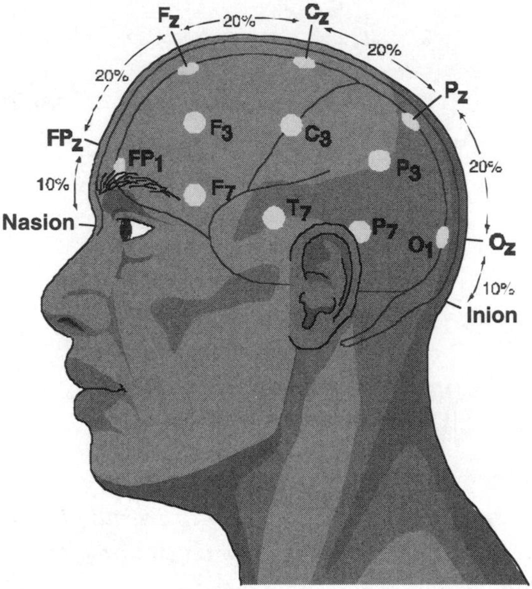

FIGURE 4.3 *Electrical activity of the brain is measured by electrodes based on a standardized system, placed on the scalp.*

mately associated with the frontal lobe. This would suggest that the empirical consciousness is a continuum and not a monothematic experience mediated by a single brain structure. The reticular activating system is biochemically heterogeneous, involving a number of neurochemicals.

Brain activity mediated by the reticular activating system is termed arousal—the physiological basis for empirical consciousness. Brain arousal is a continuum, with coma at the low end and anxiety (according to some, epileptic seizures) at the other. Measurement of the electrical activity of the brain has long been used as an index of levels of brain activation (Figure 4.3). When highly activated, the brain produces waveforms of low amplitude and high frequency (Figure 4.4).

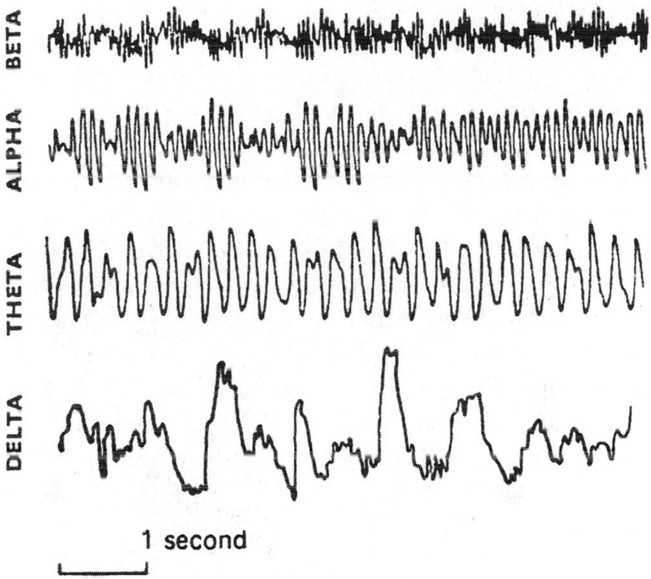

FIGURE 4.4 *Waveforms, which represent electrical activity of the brain, are conventionally divided into beta, alpha, theta, and delta, depending upon their frequency and amplitude.*

In 1949, Giuseppe Moruzzi and Horace Magoun demonstrated that electrical stimulation of the brain stem reticular activating system produced high-frequency, low-voltage electrical activity, suggesting diffuse activation of the brain as a whole. This has been confirmed with more advanced techniques.

As we have seen, the brain is made up of neurons, and like all living cells, neurons require glucose and oxygen for sustenance and activity. All expressions of neuronal activity, collective as well as individual, depend upon the energy produced by glucose and oxygen metabolism. Since the brain has a very limited capacity for storing glucose and oxygen, it is totally dependent upon its blood supply for the essential, life-giving nutrients. Cessation of brain blood circulation, even of short duration, means nonavailability of the vital glucose and oxygen. Even brief ischemia will be associated with rapid neuronal malfunction and death. Conversely, increased neural activity necessitates increased supply of glucose and oxygen and increased local cerebral blood flow (CBF).

Thus, measurement of glucose or oxygen supply and utilization could be used as indices of neuronal activity. Since blood is the vehicle through which glucose and oxygen are delivered to the brain, blood-flow measurements could also serve as an index of brain activity. Increases in brain function should be associated with increases in cerebral glucose and oxygen metabolism (CMR glucose and CMR oxygen) and cerebral blood flow (CBF), while the opposite will result in decreased brain activity. While techniques for the measurement of brain function in animals have been available for a long time, the ability to make such measurements in living human subjects is relatively new.

Close correlation has been found between blood flow to the brain stem–cerebellar region, which houses the reticular formation and states of awareness as judged by clinical and EEG evaluations in semi-coma, stupor, slow-wave sleep, drowsiness, rest, activation, dream sleep, and epileptic seizures. David Ingvar, a neurologist from Lund, Sweden, found reduced CBF and CMR in a sixty-year-old patient in a comatose state caused by an acute vascular lesion of the brain stem. There was no significant pathology elsewhere in the brain. Ingvar and U. Soderberg also demonstrated that brain stem stimulation increased arousal and CBF.

Ingvar found that when an individual is awake, blood flow to the frontal lobe is considerably higher than to the rest of the brain. This pattern of CBF was lost in a patient in a coma (secondary to a brain stem lesion). As was mentioned previously, the frontal lobe has been identified as the cortical region having the most intimate connections with the reticular activating system.

The famous Russian neuropsychologist Aleksandr Luria provided a large body of clinical and experimental evidence that implicated frontal lobes in the maintenance of arousal. Several investigators have shown an increase in frontal CBF during psychological tests.

Reduction in activation produced through tranquilization with such drugs as diazepam (Valium) reduces frontal perfusion. Similar findings have also been found in deep sleep and coma.

In summary, experiments conducted with EEG and CBF confirm the previous hypotheses that the reticular activating system of the brain stem mediates arousal and that the frontal cortex is intimately connected to the arousal mechanism (Figure 4.2).

The relationship between arousal and the subjective sense of self is highly complex. In coma, where arousal is very low, there is no subjective sense of self. In dreamless sleep, there is a sense of self, but it is vague and imprecise. In such high arousal states as anxiety, agitation, and excitement, the subject is more distinct and easily separated from the emotion. In other words, anxiety may be distinguished from the person with anxiety. Indians believe that the essential subject is separate from empirical consciousness, which arousal represents.

In Indian literature, the term "self" always includes transcendental consciousness Ātman that engulfs empirical consciousness. The lines of demarcation between the two are nonexistent, since the transcendental factor is boundless. Consciousness is the factor that makes every activity of the brain possible, and it is equated with soul. Śāṁkara states in Ātmabodha, "Ātman alone illuminates the senses and the intellect as light illuminates pots etc. [objects]" (verse 28). Ātman is the Primary Principle from which matter and the brain originated. The brain does not produce Ātman, the transcendental consciousness; Ātman produces empirical consciousness and the brain.

Ātman is the center and the hub of the mind. The *kośas* merely wrap around it—not in an anatomical sense, but in a functional sense. It is possible to reach Ātman directly. Śāṁkara points out, "When these five sheaths are taken away, what remains is the Ātman—pure, eternal, blissful, invariant, non-mediated, and self luminous" (Vivekacūḍāmaṇi, verse 151). Ātman contains consciousness, but it is in excess of consciousness and life; it is the unmanifested Absolute. The true self, according to Indian thought, is the transcendental consciousness.

Self, for most people, is the five *kośas* mixed together. The sages insist that Ātman, the transcendental consciousness, can be isolated from the rest.

This is the single most important concept in Indian philosophical thought. According to Kaṭha Upaniṣad (500 B.C.–A.D. 200): "Smaller than the small, greater than the great, the Self is set in the mind of every creature. The man at peace beholds him, freed from sorrow. Past the tranquillity of the mind and the senses, he sees the greatness of the Self" (I, 2, 20).

Consciousness: Fact or Fantasy?

Types of Brain Arousal

Neuroscientists use the term "arousal" to describe the physiologic basis for empirical consciousness. The reticular activating system, the substratum for arousal, is made up of neurons. Transcendental consciousness is the factor that endows the neurons with life, without which arousal cannot occur. Transcendental consciousness is anterior to time, space, and matter. It is one and the same as Brahman, the Absolute. Ātman is Brahman present in each individual. Since the brain is the only organ capable of introspection, it is the only site where Atman can be experienced.

This neurological explanation of the Indian concept of consciousness is tangential for a number of reasons. A number of Sanskrit words, among them *cit, prajñā, bodhaṁ, ātman, jīva, vṛti,* have been translated as meaning "consciousness." None of them denote consciousness exactly as it is understood in the West, especially by neuroscientists. While all them center around the concept of consciousness, there are subtle nuances.

Consciousness is the jagged coastline between the spiritual and the material. It is where finite emerges from the infinite, where the noumenon transmutes into phenomenon, where the Creator meets the created. It is difficult to define it with precision. (In the following chapters, the term "consciousness" is used in this imprecise sense.)

Arousal is a continuum, with coma at the low end and convulsions at the high end. The reticular activating system of the brain stem is the anatomic substrate for this physiologic phenomenon. Electrical activity of the brain recorded with an EEG best monitors arousal. Scalp electrodes pick up fluctuations in electrical activity of the brain, which vary in frequency from 1 to 50 Hz (cycles per second), and in amplitude, between 20 and 100 microvolts. Electrical signals of the brain are extremely complex and influenced by multiple factors, including signals from the brain and overlapping tissues, especially skin and muscle. Traditionally, EEG tracings are classified into alpha, beta, delta, and theta. Alpha waves are of frequency between 8 and 13 Hz, beta between 13 and 30 Hz, delta, between 0.5 and 4 Hz, and theta between 4 and 7 Hz. Alpha rhythm often accompanies calm repose during wakefulness, and it is best recorded over the posterior (parietal-occipital) brain regions. The alpha pattern of resting wakefulness is easily perturbed by such sensory stimuli as light and sound. Beta waves predominate over frontal regions, and they intensify during mental activity. Delta and theta waves accompany dreamless sleep.

There are two specific types of arousal—one that represents resting wakefulness (tonic) and a second (phasic), associated with activity. The two are closely intertwined. The famed Russian neuroscientists Ivan Pavlov and Aleksandr Luria developed the concept of cerebral tone, basic to resting wakefulness. The activity-related phenomena consist of focal changes specific to that activity, superimposed on general, more diffuse cerebral tone. Phasic events amplify tonic consciousness, at least temporarily. When someone is stuck with a pin, there is local pain but there is also an increase in general alertness. Pain is associated with increased activity in the sensory cortex and a generalized increase in arousal.

A number of tools are available to the neuroscientist to study both types of consciousness. Sudden changes in brain electrical activity associated with sensory stimulation are called evoked potentials. These responses are temporally close to stimulus induction and are very specific for the sensory system utilized. The waveforms associated with different sensory stimulations are highly characteristic and are extracted with special computerized programs that average the information from multiple stimulations. Sensory stimulation increases activity in the brain region responsible for that function. In addition, there is also a global increase with more marked changes in the frontal lobe.

The chain of structures that mediate arousal is continuous with the spinal cord below and the diencephalic structures, especially the thalamus, above. The thalamus, in turn, forms dense connections with the cerebral cortex, especially with the frontal lobe. A large body of clinical and experimental evidence, especially that provided by Luria, implicates the frontal lobe in the maintenance of cerebral tone—tonic arousal. In addition to arousal, more complicated functions closely related to wakefulness, such as the ability to abstract and reason, motivation, and planning are frontal functions. Thus, the arousal nexus seems to run through the full length of the central nervous system. In an evolutionary sense, it has intimate connections with brain regions both higher and lower. It supplies activation to all mental operations, and no mental activity can occur without assistance from the arousal chain.

For all practical purposes, arousal and consciousness may be considered to be two sides of one coin. Consciousness and arousal are paradigmatic for each and every activity of the brain. If the brain is a computer, arousal is electricity. The computer brain holds sundry programs in its meshes, but none can be activated without electricity. Electricity can be detected independent of the computer it enlivens. Can consciousness be cleaved off from the brain functions it supports? Can the background drone of the arousal chain, when the brain is not engaged in any activity, be perceived? This pivotal question, neglected by the scientist, happens to be one of the central pillars of Indian philosophical thought. As was discussed earlier, the Indian concept combines empirical and transcendental consciousness.

The Indian Concept of *Cit*

While most schools of Indian philosophy refer to consciousness in some manner, Śaṁkara has provided the most lucid descriptions.

Like the modern scientists, he accepts two types of consciousness. He recognizes the universal consciousness *(cit)*, which is the basis for all mental activities but is not directly related to any. This background level of basic consciousness is distinguished from consciousness linked to specific mental activities *(citta)*. *Cit,* which combines empirical and transcendental consciousness, is construct-free. It is not dependent upon sense organs, while sense organs are dependent upon it. *Cit* is primary to everything the brain does and, therefore, becomes the basis not only for sensations but also all voluntary and involuntary activities, both mental and physical. Thinking is dependent upon *cit,* and according to Śāṁkara, it enables the disbeliever to express disbelief in it. Consciousness is synonymous with existence, and the brain is lifeless without consciousness.

Characterizing *cit* in scientific terms is problematic for a variety of reasons. Science is built upon proof. *Cit* cannot be proven since it precedes the prover, proof, and proving. The experience of *cit* is its own proof; additional proof will be superfluous. It would be somewhat like trying to prove you are alive. Your wanting to prove it is proof that you are alive, and there is no additional need to prove it. All knowledge, including science, exists only as a predicate of *cit.*

The external world is made visible by light. The seen world disappears when light is extinguished. However, light does not have to be defined exclusively on the basis of the objects it illuminates. Light and its source can be perceived directly. Consciousness illuminates the mind. Cessation of consciousness submerges the mind into darkness. Does consciousness have to be understood exclusively on the basis of that it illuminates? Can it be perceived directly and independently?

The question plunges us into the world of subjectivity that is anathema to the scientist. Objectivity is the basis for all science. However, consciousness is the quintessential basis of the subject, and it can never be displaced into the objective world since the objective world cannot exist without it. The object simply cannot exist without the subject.

Objects are perceived, but perception is dependent upon consciousness. Thus, consciousness cannot be an object of perception. Trying to make consciousness an object is akin to including the movie projector in the picture it projects. Śāṁkara declares, "Subject and object are opposed like light and darkness." One cannot use light to examine darkness.

He also states, "Subject cannot become an object like the eye cannot see itself." This view is supported by the Upaniṣads. Kena Upaniṣad (around A.D. 200) states, "Consciousness is that which is the 'ear of the ear,' the 'mind of the mind,' the 'speech of the speech,' the 'breadth of the breadth,' and the 'eye of the eye'" (I, 2).

Intellect, the cradle for scientific thought, is not much help in explaining consciousness since consciousness antedates intelligence and therefore intelligence cannot evaluate it. Can it at least be thought of? It is impossible to know it through the medium of thought because thought itself is dependent upon it. Kena Upaniṣad explains, "[It is] not that of which the mind thinks but that by which the mind thinks" (I, 6).

Most constructs of the mind are conditioned in time and space. Perception, memorization of that which is perceived, thought that grapples with objects of perception, and language that facilitates communication concerning the empirical world are all molded in time and space. *Cit* is that which enables the empirical brain to function, but it is prior to time and space. Time and space are dependent upon consciousness and not the other way round. Transcendental consciousness is continuous with the Primary Principle, before and beyond time and space.

In the unalloyed transcendental consciousness, there are no divisions, either temporal or spatial. It is the determinate whole. It overshadows both "I" and the ego. In a realm where there is no space, "I" cannot be distinguished from "not I" since both are spatial constructs. Consciousness dissociated from mind and body becomes atemporal. Freedom from time means no changes. As we have seen, change is alteration over time, and where time does not exist, there are no changes. It does not change across the three normal phases of our existence, namely, wakefulness, dream sleep, and dreamless sleep. Where there is no time, there is no aging, where there is no aging there is no death, and that which does not die is not born. Chāndogya Upaniṣad asserts, "It does not age, it does not die with the body" (VIII, 1, 5), and Kaṭha Upaniṣad reiterates,

"The knowing Self is never born; nor does it die at any time. It is unborn, eternal, abiding, and primeval. It is not slain when the body is slain" (I, 2, 18). The essential subject remains the same while the mind grows, matures, decays, and dies.

The term "perception" cannot be used to describe subjective awareness of either empirical or transcendental consciousness. Perception, in the classical sense, involves an organ such as the eye or ear transforming the external signal into electrical impulses that the brain eventually decodes. In the subjective experience of consciousness, there is no intermediary organ of perception. It is straightforward self-realization, unassisted by an intermediary agent, and it does not involve transduction of any kind. Knowledge of the external world dependent upon the organs of perception is mediate knowledge *(vijñāna)* which is inferior. Knowledge of self, unassisted and non-mediated, is the higher knowledge, *jñāna.* Muṇḍaka Upaniṣad compares the lower and higher knowledge: "Of these the lower is the Ṛg Veda, the Yagur Veda, Sāma Veda, the Atharva Veda [scripture], Phonetics, Rituals, Grammar, Etymology, Metrics and Astrology. And the higher is that by which the Undecaying is apprehended" (I, 1, 5).

Language is constructed by, and for, the phenomenal world, and it is unable to characterize *cit,* which supersedes the phenomenal world. Śāṁkara mentions a Upaniṣadic passage in which the teacher Badva responds to his pupil Baskali's inquiry with silence. Upon repeated entreaties by the eager student, the teacher says, "I have already provided the answer, but you have not been able to comprehend. Self is silence" (Brahmasūtrabāṣya, III, 2, 17). Speech can only describe what it is not, but not what it is. Bṛhadāranyaka Upaniṣad (before 500 B.C.) simply says, "Neti, Neti," "not this, not this" (IV, 2, 4). Years ago, I came across a poem on the resurrected Lazarus's futile attempts at describing his experience of death. He, too, was only able to say "Neti, Neti," in Hebrew.

Levels of Consciousness

The vestures of the mind previously considered are fused with consciousness, and they are usually perceived together. This is true of the available neurophysiological research on consciousness. Most articles on the physiology of consciousness are on such brain activities as perception and cognition. Śaṁkara compares this relationship between consciousness and various functions of the mind to the relationship between the light and a crystal placed in front of the light; the light takes on the color of the crystal. An onlooker may conclude that the color of the crystal is the color of the light.

According to the Taittrīya Upaniṣad, the veils that cover the consciousness represent different stages of evolution. The cosmic cycle starts with the Absolute, which is synonymous with transcendental consciousness. Matter, or *anna*, is derived from it. Matter gives birth to life or empirical consciousness. Over time, life expands to form the mind. The mind develops the ability to acquire knowledge that takes it to the next phase of *vijnāna*. First, the less-sophisticated empirical knowledge regarding the external world is acquired. Then the capacity to turn attention inward to acquire the superior knowledge of self is gained.

Consciousness, at the most basic level, transcends time and space and becomes independent of the matter of which the brain is composed. Since evolution of the mind represents progressive superimposition of created elements of increasing complexity upon consciousness, the more advanced the organism in the chain of evolution, the more contaminated consciousness will be. Thus, a dog will manifest the Primary Principle more clearly than the Nobel laureate. Perhaps that is why some of us envy animals and consider associations with them blissful. The purity and simplicity they display belie proximity to the Absolute.

The arousal chain that mediates empirical consciousness is not monolithic; it is made up of the core structure brain-stem reticular formation, the thalamus, and the frontal lobe. According to the well-known British neurologist Hughlings Jackson (1835–1911), different levels of brain organization represent rungs of the evolutionary ladder. At the lowest level, the spinal cord, the medulla, and the pons carry out less-differentiated functions essential for the preservation of life. The sensory, cognitive, and conative apparatus represents the next level. More functional differentiation is evident here. According to Jackson, the frontal cortex signifies the highest level of brain development, where the greatest functional differentiation, specialization, integration, and cooperation are seen. If Jackson was correct, and I do believe he was, consciousness becomes more differentiated as it wafts up from the brain stem to the frontal cortex. It also means that organisms that occupy different levels in the scale of evolution will differ in the architectonic of arousal, and therefore their experience of empirical consciousness will not be identical. Transcendent consciousness is undifferentiated and the same for all.

It needs to be noted that even the most primitive organism is a created element and not totally transcendent. The very act of creation involves disguising the Primary Principle in *māyā*. This, however, will be less in a primitive organism. In spite of their proximity to the Primary Principle, animals do not have the ability to transcend the empirical components of their minds through volition. Vedic literature points out that only humans can accomplish total transcendence through effort. Taittrīya Upaniṣad alludes to this. While considering different explanations for consciousness, the teacher points out that simple instinctual experience of life (which animals have) does not represent the deeper levels of consciousness humans can access.

Desire requires the ability to conceptualize the desired. We do not wish for something we are totally unaware of. Humans desire enlightenment because they are able to glimpse it or intuit it. To realize the desire, effort needs to be expended. This may involve activation or deactivation of different brain mechanisms. For example, we now know that falling asleep involves activation of a sleep center, which in turn deactivates most of the brain. Only humans have the ability to intuit enlightenment, to seek it, and to manipulate brain mechanisms to realize it. Partial proof for this comes from the well-established finding that animals do not like such drugs as LSD, mescaline, and marijuana, which have close associations with spiritual states of mind.

This would indicate that the more recently evolved brain structures, especially the frontal lobe, are necessary to overcome the time-space barriers and to reach transcendental consciousness. The frontal lobe sets targets, develops a plan of action to get there, and gives effect to the plan. It is also responsible for staying focused, both during the planning and executive phases.

What proof is there for these far-fetched claims? The argument made by Indian philosophers that the unformed rudimentary consciousness is transcendental has their personal experience as its exclusive basis. Subjective, and less reliable though this may be, it cannot and should not be dismissed lightly for at least two reasons. First, it will be difficult, if not impossible, to verify this experimentally for reasons already given. Second, a chorus of voices spanning the entire length of Indian philosophical thought, covering many generations over a period of 2,500 years, supports this claim.

The brain is responsible for all cognitive, conative, and intellectual functions. Transcendental consciousness, however, is a striking exception. It is not a function of the brain, it does not reside in the brain, and the brain is not required for its existence. It is necessary for the existence of the brain and everything it does. The idea that a nonmaterial, ethereal entity called consciousness may supersede the brain may seem chimerical. Neuroscientists may frown upon this notion and reject it out of hand so long as they believe in absolute time and space and the supremacy of matter, as physicists of the last century did. According to Einstein, "In this new kind of physics there is no room for both matter and field, for the field is the only reality."

The scientific community acknowledges that there is an Absolute before and beyond the slippery slopes of time and space. Scientists also agree that this is beyond human logic and its products, including science and mathematics. What the Advaita philosophers argue is that while it clearly is beyond human logic, unfettered application of logic can take you to the very edges of reality, for which physics and mathematics have provided proof. However, only extralogical intuitive insights can carry us further. They also believe that the access to the Primary Principle exists in the unfathomed depths of the human mind.

Humans are part and parcel of the universe—and whatever of which, and by which, the world is composed, will have to be the ultimate fabric and foundation for humans as well. The elegant equations, which describe the mysteries of the cosmos, will also have to explicate human life and consciousness. There cannot be one Absolute for the cosmos and another for the human mind. There cannot be one truth for the physicist and another for the neuroscientist. What is true for the macrocosm has to be true for the microcosm as well. As physicist John Wheeler of Texas has averred, the search for eternal principles will have to include the experimenter in the experiment, for without the experimenter the experiment will be incomplete. Study of the object will yield only incomplete and partial solutions if the subject, without which the object cannot exist, is excluded. Mind is a projection of the brain, and brain a creation of matter, and matter an expression of the Absolute Principle.

Absolute Consciousness and the Divine

Consciousness is the spirit within a material frame. It is the unchanging ground for the changing formation of the manifested world, much like the windblown wisps of clouds against the steady, shimmering moon. In the absence of consciousness the created world has no existence, and therefore it may be argued that consciousness is the creator and the destroyer.

Absolute consciousness in Sanskrit is referred to as Brahman, the Primary Principle, God. Brahman present in each living organism is called Ātman. They are one and the same. Śāṁkara condensed his entire philosophy in two sentences in Brahmajñānāvalīmālā (20):

> *Brahman is real, the world is not*
> *Consciousness and Brahman are not different.*

The notion that the divine is contained within the human mind is not unique to India. People of all religious persuasions prefer quiet, tranquil environments when they pray. Most close their eyes, which would suggest they believe God is inside and not outside. God can be reached anywhere and everywhere, on earth, in water, in air or space. That which accompanies you wherever you go is your mind. Thus, it would also indicate that God is in your mind. The Gospel according to Luke speaks to the point: "The kingdom of God cometh not with observation. Neither shall they say, Lo here, lo there! For the kingdom of heaven is within you" (17.20–21). The same view is expressed by John and Paul as well. "The kingdom of heaven is the highest state attainable by man. It is within us" (John 3.3). "Know you not that you are the temple of God and that the spirit of God dwelleth in you." (Corinthians 3.16)

There have been moments in the lives of all thinking people when we doubted the reality of the perceived world, searched for the fact concealed by the fiction, and pondered the true meaning of existence. Reports of the senses may be deceptive and memory based on them may be sheer fantasy. The great curve of time from the past to the future may be a figment of our imagination. Names and forms, which characterize the manifested world, may be nothing but dancing shadows. However, amid the fogs of uncertainty, we are certain of the reality of our own existence, and as Śāṁkara points out, few people argue that they are unreal. Our own existence, even when cleaved from the world that surrounds us, upon which doubts are cast, escapes the thorns of doubt. The lifelong journey in search of truth comes to an end in the inner sanctum of our own minds. All streams of consciousness and tracts of experiences commence and terminate with the self. Śāṁkara exemplifies the point with the story of the woman who desperately searched for her precious necklace without realizing it was around her neck all the time.

Among the schools of Indian philosophy that contradict this viewpoint, Buddhism stands out. According to it there is no permanent self at all. Some conservative Buddhists—Theravādins—hold that the conscious experience is simply a series of momentary experiences strung together by the thread of time. However, this theory, known as Śūnyavāda (the emptiness hypothesis), does not explain the self—who witnesses and experiences the emptiness. In addition, in order for one to coalesce the free-standing series of momentary existences into a continuous whole, one has to have a memory store that maintains memories of the past so that they may be merged with experiences of the present. Once again, this requires an entity with memory of the past, experiences of the present, and the ability to fuse the two. The reformed Buddhists—Mahāyānins—interpret Śūnyavāda differently.

The Śūnyavāda can be traced back to the Buddha, who declared that nothing was permanent, including consciousness. He held that all these were fleeting fantasies lacking in substance. All desires based on these fantasies were also considered to be empty and their pursuit, foolish. Life is a stream of creation and destruction, whether it lasts a split second or a thousand years. The Buddha's understanding of the manifested world appeared to be compatible with the Yin-Yang principle. He argued that the world has duality as its basis, consisting of "it is" and "it is not," and such a dichotomous world cannot be real. The world, according to the Buddha, is like a glowing stick that when whirled around creates the illusion of a circle. He also considered body, mind, and intelligence as part of the illusory whirl. The manifested world appears and disappears in rapid succession, but the speed is considerably faster for the mind than it is for the body and the external world. He used the wheel mentioned earlier to illustrate the relationships between time and perceptual reality. The wheel of life moves us forward into the future from past; the present is the point of contact between the wheel and the ground. The whirl of the glowing stick and rolling of the wheel are not random events—the cosmic law, or Dharma, governs them.

Statements about unreality can make sense only in comparison and contrast to reality. While the Buddha spoke extensively about the phantasmal nature of existence and phenomena related to it, he did not say a whole lot about reality. Not commenting on the element of permanence against which the ever-changing flux of manifesting and becoming could be gauged meant that his arguments left a number of loose ends. While the Upaniṣads posited an unchanging Absolute Reality around which the evanescent changes in time and space occurred, Buddhism was silent or noncommittal about this.

Emptiness, or *śūnyatā,* comes close to the Buddhist concept of reality. As noted, the Buddha did not provide detailed explanations. But later Buddhist writers, especially, Nāgārjuna, the second-century Buddhist philosopher, wrote extensively about it. According to Madhyāmaka, his system of philosophy, the phenomenal world is relative to causes and conditions *(pratītiyasamutpāda)* and it has no separate reality *(svabhāva)*. It is a card castle that has meaning and substance only when arranged in the proper sequence. The whole edifice will collapse if a single card is taken out. Nāgārjuna denies even independent existence for the individual cards—each has existence only when arranged as a castle. In many ways, Nāgārjuna's explanation of the phenomenal world is close to Einstein's theory of relativity. The insubstantial nature of the manifested world is synonymous with *śūnyatā*. Although the word has been translated as nothingness or emptiness, it does not have an equivalent in English. Etymologically, it is derived from the Sanskrit root word *śvi*—to swell, to expand—very close to the word *brh* from which the Upaniṣadic Brahman is derived. *Śūnyatā* is the nothingness pregnant with the potential for everything.

There are any number of reasons why the Buddha may have chosen to remain silent on this issue. The Buddha was driven by his compassion for human suffering and not by the desire to unlock the mysteries of the cosmos, the secret of life, and the true meaning of creation and destruction. His primary goal was to find a solution for the pain and misery that plagued human existence. The Buddha argued that if somebody was lethally wounded with an arrow, first and foremost, the arrow must be extracted and only then should one concern oneself with where the arrow came from.

The intellectual doyens and luminaries of the time took special pleasure in sporting with metaphysical conundrums. Neither the dialogues and debates the intelligentsia engaged in nor the sacraments and sacrifices the priestly class practiced nor the self-denial and mortification the recluses submitted to held much meaning or substance for the average man in his struggle with living and dying. The Buddha had tried his hand both at self-mortification and intellectualization before he chose the path that eventually led to his enlightenment.

The religion he sought to found was unstained by religious dogma, superstition, or priest craft. Freedom was its central stronghold, and every person was as important as every other, regardless of social standing, material prosperity, or position in the caste hierarchy. His religion was for everyone, not just the intellectual giants. The Buddha was particularly reluctant to engage in discussions and debates over the incomprehensible and the unfathomable. To him, both the problem and the solution were unique to each individual. It depended on each person's world view, philosophy of life, and conduct. You are because of what you have been.

The Buddha was very cognizant of the fact that he had left a number of metaphysical issues open. He also knew that the answers to a number of these issues were beyond human comprehension. Once, when the Buddha was in the forest with his disciples, he grabbed a handful of leaves and asked his followers whether there were more leaves in his hand or on the ground. Unanimously, all of them said that there were definitely more on the ground. The Buddha declared that the material to be taught and learned was represented by the leaves in his hand. But he acknowledged that there were many more truths as represented by the countless leaves on the forest floor.

The Upaniṣads, on the other hand, did attempt to tackle these issues, although they, too, acknowledged the futility of the task. For example, they admitted the ineffability of the pure conscious experience. While they claimed silence was the best way to describe it, they attempted to provide some idea, though partial and incomplete, of what it was through the medium of language. It is conceivable that the Buddha accepted the ineffability, in the full sense, and did not attempt to describe the state at all. He probably felt that the only way to know was to experience it. Descriptions would be a waste.

Later Buddhist writers characterized empirical consciousness as *vijnāna*, and the underlying transcendental consciousness as *nirvāṇa.* However, the Buddha defined *nirvāṇa* in negative terms only: the passing away so that no passion remains, the giving up, the getting rid of, and the emancipating from. It is a state of being totally devoid of ego and related attachments, hopes, ambitions, and desires.

In order for the Buddha to extol emptiness and for the Vedic seers to glorify consciousness, they must have had at least some experience of it. We are told that the only proof is personal experience. We can either take the masters' word for it or go in for a personal experience. It would also seem that in order to disentangle consciousness from other activities of the mind, we should have special abilities or engage in esoteric practices. On the other hand, if consciousness is the ground for the mind, it has to be present in all phases of conscious experience and must be accessible at some point. We may not be able to view the sun directly, but we can have some idea of it from the objects that reflect its light. If consciousness is the sum and substance of life, every living person must have had at least a passing glance at it.

The Phases of the Mind

In Indian philosophy, overlying veils of the mind distort pure consciousness so that under ordinary conditions it is inaccessible. It is disguised as three states, namely, sleep, dream, and wakefulness. Everything we experience, memorize, know, and learn is dependent upon one or more of these three conditions. In fact, the story of our entire life is contained in these three chapters. Even people with just a nodding acquaintance of Indian philosophy are familiar with the expression "Aum," or "Om." It is not widely known that "Om" and the English prefix "omni," meaning "all," may have the same etymological roots. "Aum" stands for the totality of human experience, every single thing we can perceive, intuit, or infer. It is emblematic of the creator and the creations, which in Indian thought are one and the same—Brahman. A very old word going all the way back to the Vedas, before the Upaniṣads, it is first mentioned in the Black Yajur Veda, composed sometime between 1000 and 500 B.C. It is also mentioned in the Brāhmaṇas (parts of Vedas that deal with liturgy), as a response to the Ṛg Vedic verses uttered by the priest, in a manner similar to the use and meaning of "Amen" in Christian ceremonies. The term became the symbol representing the universe in Aitareya Brāhmaṇa (verse 32). The waking state is the letter "A", the dream state is the letter "U", and the state of deep sleep is the letter "M" (Māṇḍūkya Upaniṣad, 9–11).

We spend at least one-third of our lives in sleep. Different from coma, sleep can be reversed through strong sensory stimulation. All living creatures, including single-celled ones, show some rudimentary form of sleep, mostly as diurnal variations in rest and activity. As the complexity of the brain increases, the complexity of sleep shows parallel changes. On the evolutionary scale, dreaming appeared later than non-dream sleep. Animals with less developed mental capacities, like dogs and cats, dream in the same way as humans; therefore, the argument that dreaming is related to psychological factors unique to humans does not have much scientific support.

The ability to record electrical activity of the brain with scalp electrodes ushered in a new era in sleep research. The noninvasive electroencephalograph made it possible to study electrical activity of the brain in large numbers of subjects while they were asleep. This information enabled researchers to divide normal sleep into four phases. Under wakefulness, an EEG will show very fast, low-voltage electrical activity. The activity is saltatory and desynchronized, indicating uncoordinated discharge of many neurons. Under resting conditions with eyes closed to exclude visual input, the frenetic activity of the wakeful brain becomes slower and better synchronized, with the voltage becoming higher and the frequency lower (Figure 5.1). Low-voltage (less than 50 microvolts), fast (15–50 cycles/second) activity is replaced under resting conditions with high amplitude and slower and smoother waves, usually referred to as alpha waves (8–12 cycles/second). In general, as sleep deepens, the EEG becomes progressively slower in frequency and deeper in amplitude. Stage 1 sleep is usually associated with 4–8 cycles/second frequency, Stage 2 with 8–15 cycles/second, Stage 3 with 2–4 cycles/second, and Stage 4 with 0.5–2 cycles/second. This cycle starts with Stage 1 and in about 90–100 minutes finishes Stage 4 to return to Stage 1 again. Thus, the Stages 1–4 sequence repeats itself every 90–100 minutes (Figure 5.2). Stage 1 is generally regarded as the pre-sleep stage. Stage 2 shows definite evidence of sleep onset.

Continued research along these lines revealed phases during which eyes darted from side to side and the EEG resembled normal wakefulness, with high frequency, low amplitude tracings, although the

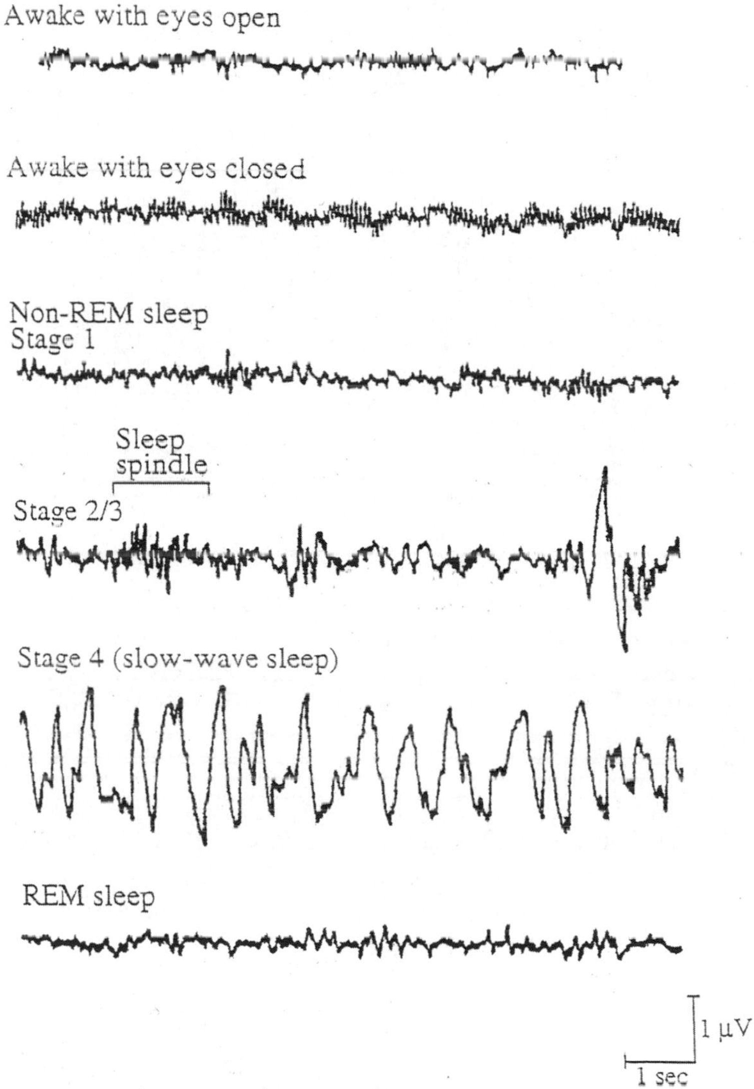

FIGURE 5.1 *As sleep deepens, EEG becomes progressively slower in frequency and deeper in amplitude. REM sleep, or dream sleep, is associated with brain electrical activity similar to that of wakefulness with eyes open.*

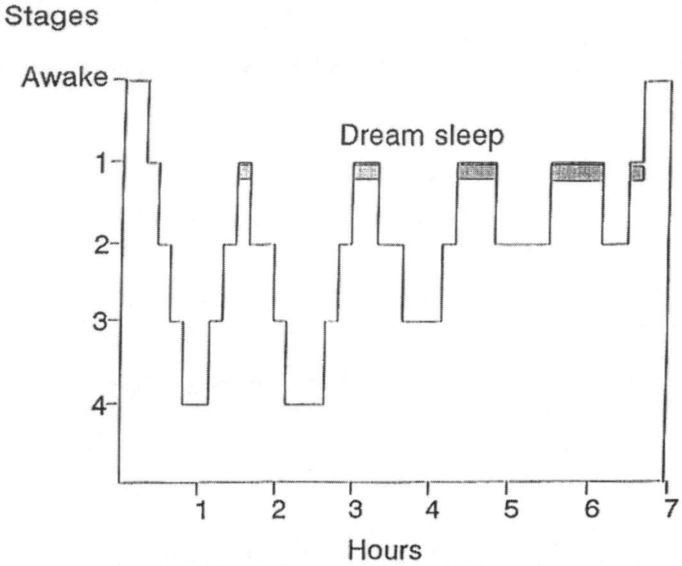

FIGURE 5.2 *Sleep occurs in cycles lasting 90 to 100 minutes, representing the transition from Stage 1 to Stage 4 through the intermediary steps. After Stage 4, sleep lightens and returns to Stage 1, when REM, or dream phase, will set in. The next cycle follows this, and so on.*

subject was still asleep. These phases were dubbed REM sleep (rapid eye movement sleep), or paradoxical sleep. When awakened from REM sleep, subjects reported vivid dreams. Most of the time, when awakened from non-REM sleep, subjects did not report dreams, and if they did, they were unlike classic dream experiences. Increased respiration, pulse, and decreased muscle tone accompanied REM sleep. REM takes place only during Stage 1 sleep. Normal adults spend approximately one-fourth of their sleep time dreaming.

Sleep occurs in cycles lasting 90–100 minutes, representing the transition from Stage 1 to Stage 4 through the intermediary steps. After Stage 4, sleep lightens and returns to stage 1, when REM phase will set in. The next cycle follows this, and so on. The duration of the REM phase increases as the night progresses, and most of the REM sleep occurs during the second half of the night. Non-dream sleep, on the other hand, occurs mostly during the first half of the night. Stages 3 and 4, representing the deepest sleep, usually occur during the first sleep cycle. Individuals awakened from Stage 3 or Stage 4 sleep are usually confused and disoriented for several minutes. As the night progresses, the last two sleep cycles may have no Stage 3 or Stage 4 (Figure 5.2).

Wakefulness, as we have seen, is associated with high-frequency, low-voltage activity of the EEG, indicating burgeoning activity of multiple groups of neurons. Wakefulness, of course, is associated with increased basic levels of brain activation. Superimposed on that are the activities of neural populations representing the multifarious functions of the brain, conscious and unconscious, in the areas of cognition, conation, and mentation. Waking brain activity, to a considerable extent, is dependent upon and related to the created world. Perception, cognition, memory, thinking, motivation, and action are all closely intertwined with the manifested world, conditioned in time and space.

During REM sleep, electrical activity of the brain is similar to wakefulness with fast low-voltage activity. The reticular formation in the pons is powerfully stimulated, actually more powerfully during dream than during wakefulness, and the brain is aroused as it is during wakefulness. Cells of the visual cortex are just as active during REM sleep as during waking, which might explain vivid visual images associated with dreams. Other sensory systems have also been found to be activated during REM sleep. The thalamus, which communicates sensory information from the periphery to the cortex, including sense of balance and position, is also active during REM.

Since the brain produces dreams entirely on its own, with no external sensory data, the dream content has to have both psychological and physiological significance. A number of individuals have attempted to provide explanations for dreams and dreaming. Sigmund Freud believed dreams are concealed efforts at primitive wish-fulfillment. According to him, the human mind harbors crude bestial instincts with incestuous, patricidal, and cannibalistic passions constantly trying to force their way into conscious awareness in pursuit of fulfillment. The conscience-laden superego forces them back into the dark depths of the unconscious. Dreaming is a mechanism that allows partial fulfillment of these reprehensible and repulsive desires in a concealed and distorted way. The process of concealment, which involves a number of devices, including symbolization, is referred to as dream work.

Advaita philosophers have not said anything that would support or contradict Freud's postulates, which he formulated at least a thousand years after Advaita. They point out that in dreams the mind is able to create everything perceived and experienced during wakefulness. All sensory modalities are active and thinking; willing and acting take place in a manner comparable to wakefulness. The dreamer is totally convinced of the reality of the dream experience. Wakefulness sublates dream; upon waking up, the dreamer concludes that the waking experience is more real than the dreaming one. If we never woke up from a dream, we would never realize that we were dreaming. The Advaita position is that what is experienced during wakefulness is also unreal. Advaita philosophers hold that divisions along the dimensions of time and space necessary for the creation of the manifested world are a distortion of the transcendental reality. The same mechanism operates in dreams. The brain is still in a mode wrought in time and space, and images brought up in dreams, whether from the past or future, are still cast in the same time-space mold into people and things. So, the mechanism of *māyā* that is there during wakefulness is still there, albeit with some loosening. Dreams associated with REM are very vivid and emotional but also very bizarre. Spatial and temporal rules of the manifested reality are often violated. Scenes switch rapidly, sizes and shapes alter, deceased people are brought back to life. The dreamer is transported back in time to the distant past or propelled into the future.

Thus, although there are differences between waking and dreaming, for all practical purposes, the differences are insignificant. Śaṁkara explains: "In the dream state where there is no object, mind on its own creates everything; and it also creates everything seen when awake as well. There is little difference between the two states. Everything we see is a creation of the mind" (Vivekacūḍāmaṇi, 170.)

Both Śāṁkara and the Buddha recognized and accepted the precognitive element of dreams as an expression of time transcendence. Śāṁkara acknowledged the desire-fulfilling characteristic of dreams, but he also regarded wakefulness as desire-based. The Buddha also emphasized the important role of desires in creation of normal wakefulness. Neither Śāṁkara nor the Buddha differentiated between sordid and decent desires as Freud did. Both decried the frenzied chase after imagined goals and averred that the more fruitful goal would be to search for the invariant reality. Jesus also propounded the same message, about 500 years after the Buddha and 800 years before Śāṁkara.

Dreamless sleep is experientially, anatomically, and physiologically distinct from wakefulness and dream sleep. Non-dream sleep occupies approximately 75 percent of sleep time. During this phase, the brain, for the most part, and the body go into relative inactivity, presumably to recuperate. Body metabolism slows down, and heart and respiratory rates decline steeply, especially in Stage 4. Brain blood flow and metabolism have been known to drop dramatically during Stage 3 and 4 of dreamless sleep (Figure 5.3). In other words, this phase of consciousness is one in which the brain turns off its internal lights, except for a few in the engine room. All brain activities cease, except for the few necessary to sustain life. As we have seen, progressive slowing of the brain's electrical activity reflects this.

Empirical faculties of the mind are inactive during dreamless sleep. If there is such a thing as experience of uncontaminated consciousness, the experience during dreamless sleep has to be very close to it. Mental activities related to the material world, including perception, cognition, and conation are suspended during dreamless sleep. Thus, the underlying consciousness, unsullied by *māyā,* shines through. In deep sleep, time and space constraints imposed by the empirical world, as is the case with waking and

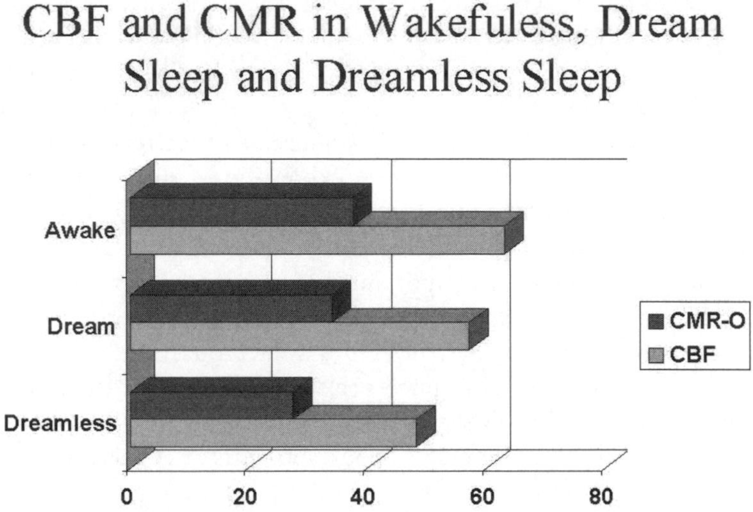

FIGURE 5.3 *Brain blood flow and metabolism have been known to drop dramatically during Stages 3 and 4 of dreamless sleep.*

dreaming, are removed. Material desires, which dictate and determine waking and dream experiences, are also absent. Yin-Yang distinctions, which characterize waking and dreaming, are suspended during dreamless sleep. According to Śāṁkara, "During dreamless sleep when there is no object at all, no pleasure or pain, consciousness shines in its own glory" (Vivekacūḍāmaṇi, 107). Bonding of the consciousness with body and mind, which gives rise to such false notions as "I am my body," and "I am my mind," are lost. What remains in the dreamless sleep is less contaminated uncontaminated consciousness, which is the true self.

Śāṁkara argues that consciousness is present in the three states of consciousness, however it is heavily contaminated by the phenomenal world and related mental mechanisms in the first two. Any difference in the feeling of self one perceives between the three states is entirely and exclusively due to this contamination and not due to any changes in the underlying consciousness as such. Śāṁkara makes this point very clear in verse 125 of Vivekacūḍāmaṇi, "There is one—self—existent, eternal, who is the base of all beliefs that 'I am,' and who is the witness of the three states of consciousness—waking, dreaming, and dreamless sleep—and who is separate from the five sheaths [of the mind]." While it is true that contamination by *maya* is minimal during dreamless sleep, it is incorrect to state that it is totally absent.

Dreamless sleep is definitely not a negative experience. In fact, most if not all, individuals consider it a pleasurable experience. Many eagerly look forward to it. Most are displeased when forced to wake up and get out of bed. This aspect of the dreamless sleep has escaped scientific attention, if not scrutiny. No explanations have been put forward for the "reward" attached to dreamless sleep.

The Upaniṣads pay special attention to this very unique pleasure, for in their view it has special spiritual significance. This is the *ānandamaya kośa* revealed during the slumber of the overlying *kośas*.

Deep dreamless sleep, according to Indian thought, is a window into the non-dual, undifferentiated consciousness. In Bṛhadāranyaka Upaniṣad, sage Yājñavalkya, in his dialogue with King Janaka, says: "[In deep sleep] he becomes transparent like water, without quality. This is the world of Brahma, Your Majesty" (IV, 3, 32). It should be noted that deep dreamless sleep is not total transcendence; among the three phases wakefulness, dream, and dreamless, it is the closest.

Deep sleep has to be differentiated from unconsciousness and coma. Coma is experientially, physiologically, and philosophically distinct from sleep; the brain is even less active. Electrical activity of the brain is considerably slower in the comatose patient than deep sleep, and cerebral metabolism (Figure 5.4) and blood flow also show very low values. Death is associated with electrical silence of the brain and absent cerebral blood flow and metabolism (Figure 5.5). Thus, deep coma and unconsciousness resemble death more than normality.

According to Paiṅgala Upaniṣad, "Unconsciousness resembles the state of a dead man" (II, 10).

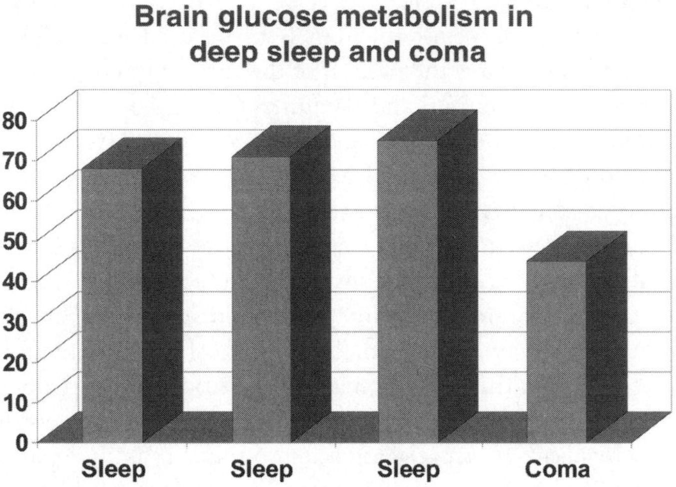

FIGURE 5.4 *During coma, cerebral glucose metabolism is considerably lower than during dreamless sleep.*

SOURCE: W. D. Obrist, T. A. Jennarelli, H. Segawa, C. A. Dolinskas, and T. W. Lingfitt, "Relation of Cerebral Blood Flow to Neurological Status and Outcome in Head-Injured Patients," *Journal of Neurosurgery* 51 (1979):292–300.

The argument may be made that the real should transcend death as well, since birth and death are phenomenal and unreal. Thus, coma should facilitate transcendence even better than dreamless sleep. If the brain is not essential for transcendental consciousness, then coma (death) should take us closer to it than deep sleep.

According to Indian thinkers, however, death does not automatically mean transcendence and experience of the unsullied Absolute. Individuals firmly yoked to the finite world will continue their enslavement to it, even after death, through reincarnation. The interesting concepts of death, rebirth, and reincarnation shared by the Upaniṣads (Paiṅgala Upaniṣad, II, 11) and the Buddhists are unfortunately beyond the scope of this book.

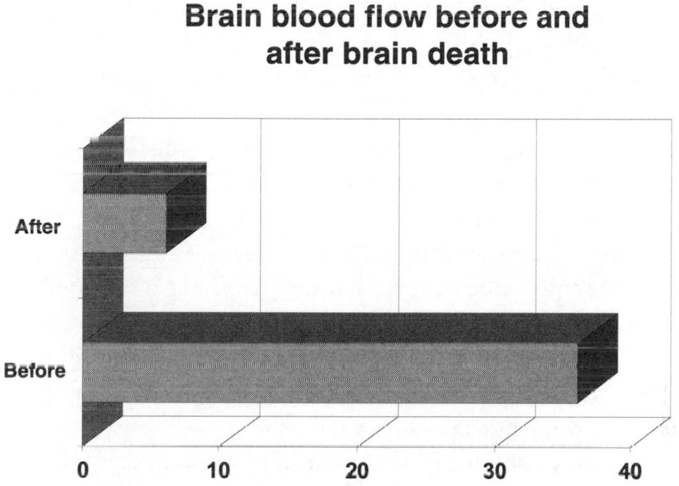

FIGURE 5.5 ***Brain blood flow before and after brain death.*** *Brain blood flow is very low or absent during brain death.*

SOURCE: W. D. Obrist, T. A. Jennarelli, H. Segawa, C. A. Dolinskas, and T. W. Lingfitt, "Relation of Cerebral Blood Flow to Neurological Status and Outcome in Head-Injured Patients," *Journal of Neurosurgery* 51 (1979):292–300.

The brain generates three different states of awareness: wakefulness, dream sleep, and dreamless sleep. The first two are dominated by time-space distinctions and, therefore, regarded as unreal by the ancients. This notion, as we have seen, has received oblique support from modern neuroscientists. Of the three states of mind, the dreamless sleep bereft of time and space is regarded as closest to reality by the Upaniṣads. Here the superficial vestures of the mind, namely the *annamaya kośa, prāṇamaya kośa, manomaya kośa,* and *vijñānamaya kośa* are rendered dysfunctional.

From studies of cerebral blood flow and metabolism, we know that brain regions that mediate these functions become inactive during this phase. The bliss experienced during deep sleep is due to the undisguised *ānandamaya kośa.* While dreamless sleep is very close to the Primary Principle, it is not quite there.

Although deep sleep may provide a blissful access to the universal consciousness, only a few wake up in the morning enlightened. Indian thinkers recognize a fourth state, Turīya, or Chathurtha, the total absorption into absolute transcendental consciousness without even the faintest shadow of duality. The word *turiam* means "the fourth." *Turīya,* devoid of even the *ānandamaya kośa,* is the true beyond. It is the beatific vision of reality, transpersonal intelligence, and eternal wellspring of knowledge (Māṇḍūkya Upaniṣad, verse 7).

This state has no sense of self, no ego and no subject-object distinction, even faintly. Since it is totally independent of the phenomenal world, language cannot describe it.

Wakefulness sublates both dream and dreamless sleep. Turīya sublates everything else. It enlightens the aspirant to the ultimate unity of things and merges us with the primordial Absolute Principle. Once having been exposed to Turīya, we are convinced beyond the shadow of a doubt about the unreality of the waking experience, the same as when upon waking up, we gain insight into the unreality of the dreaming experience.

Śāṁkara writes, "Dispelling all the distinctions that are created by *māyā,* it ever exists—unchangeable, without parts, not to be measured, without form, without name, endless, self-luminous and independent" (Vivekacūḍāmaṇi, 238).

They hold that this state is beyond and before the brain. The brain is derived from it and not the other way around.

Pharmacological Spirituality

Early Drug Experimentation

Unlike their animal predecessors, primitive humans with expanded intellect were not willing to accept their lot, to maintain the status quo. They began to explore and experiment.

Nourishment upon which life hinged was not always easy to find. Foraging for food involved ingesting different plants, fruits, and vegetables, some of which served as food, while others relieved various ailments. Still others changed the experience of self and surroundings. Some produced ecstatic mood states, and others opened new vistas of knowledge and experience. These drugs and their usage occupied an exalted position in early societies, including those of India.

It is generally accepted that Ayurveda, the Indian indigenous system of medicine, had its humble origins there, before the second millennium B.C. The Āryan hordes that came over the Himalayas brought with them not only their Vedas, but a sacred drug, soma.

Soma, which purportedly catalyzed communication between humans and Gods, was deified. All Vedas make references to it. More than 100 hymns from the anthology Ṛg Veda are about soma and its usage. Soma was a vegetal derivative, and Ṛg Veda provided details of pressing the plant between stones in wooden bowls and extracting it via filtration with wisps of wool. Use of soma was enshrouded in elaborate ceremonies, and rigid rules controlled who consumed it and how. It was probably mixed in water or milk. Of the ecstasy induced by the drug it is said, "Where there are joys and pleasures, gladness and delight, where the desires of desire are fulfilled, there make me immortal" (Ṛg Veda 9, 113). Soma raised the devotee to dizzying heights of spirituality.

One Ṛg Veda verse says: "We have drunk the soma; we have become immortal; we have gone to the light; we have found the Gods. What can hatred and the malice of a mortal do to us now?" (verse 8, 48).

Apparently, the drug also produced unpleasant reactions, reminiscent of the bad trips following the ingestion of modern drugs such as LSD. Ṛg Veda implores of king soma: "Have mercy on us for our well-being. Know that we are devoted to your loss. Passion and fury are stirred up. Oh, drop of soma, do not hand us over to the pleasure of the enemy" (8, 79).

Although the Upaniṣads and Bhagavad Gītā, of a later date, mention soma in reverential terms, with time the ritual lost its appeal and popularity. According to *Buddhacarita* (Acts of the Buddha) by Aśvaghoṣa, the second-century chronicler, young Siddhārtha, before he became the Buddha, drank soma and experienced the bliss of spiritual tranquillity (ii, 37). In present-day India it is seldom, if ever, used, and only a few people even know what soma is.

After much speculation by various scientists, R. Gordon Wasson of the Botanical Museum of Harvard University identified the mythical soma as *Amanita muscaria.* Ibotinic acid is the active ingredient in the *Amanita muscaria* mushroom. Also known as fly agaric—it intoxicates and induces sleep in flies—it is a red mushroom speckled with white.

Soma use was by no means restricted to the Āryans who came to India. References to "hoema" can be found in the Avesta of ancient Persia. The drug was also associated with religious ceremonies among the Koryak tribe in Siberia and by Native American tribes who lived around the Great Lakes. Since nomads from Asia who crossed over the region of the Bering Strait may have settled the Americas, it is conceivable that the sacramental use of the mushroom may also have spread to the Americas from Asia.

A number of other vegetal derivatives were used in different parts of the world for their effects on the brain. More than 3,000 years ago, the Aztecs used Mexican mushrooms in their ceremonics. The Huichol of Mexico turned to peyote. After using *ayahuasca*, South Americans concluded that the drug-induced state was reality and that waking conscious experience was an empty play of fantasy. Among the drugs used to enhance spiritual ecstasy were the water lily *(Nymphaea caerulea)*, mandrake *(Mandragora officinarum)*, the opium poppy *(Papaver somniferum)*, morning glory *(Turbina corymbrosa)*, parican *(Genus virola)*, ibogaine *(Tabernanthe iboga)*, and *Cannabis sativa*.

These drugs with potent effects on the brain and behavior were put to both benevolent and malevolent uses in different parts of the world. Medieval witches of Europe are believed to have used members of the nightshade family in their broths and potions. Initiation to join secret societies and certain cults in Gabon and the Congo involve ingestion of iboga, and witch doctors establish communication with the spirit world through that drug. There is no question that pharmacological agents have been utilized all over the world to produce alterations in the mental condition deemed supranormal, extrasensory, and profound by the users.

Since drugs act upon the brain to bring about their effects, study of these drugs is likely to reveal brain mechanisms related to spirituality. If drugs enhance or induce spiritual experiences, it would seem highly likely that such experiences are mediated through neural mechanisms.

Unfortunately, such research is not easy. Of the 150 or so such drugs known to Western science, very little pharmacological information is available on most. Many plants used in religious ceremonies contain chemicals other than the substance responsible for the spiritual effects, and their effects, too, have to be taken into account. Several of these plant extracts are too toxic to be administered to human volunteers. Identification, isolation, and purification of the active ingredient are by no means easy and have not been carried out in many cases. Animal studies are obviously easier, but it is unclear whether animals are able to have spiritual feelings, and even if they are, they will have difficulty communicating such to the experimenter.

Animals do not seem to care for many drugs used for consciousness alteration in humans, such as peyote and psilocybin, which would suggest brain mechanisms unique to humans. As was discussed previously, ancient Indians believed only humans can access supernal states of mind. However, we do have access to a few compounds safe for human administration. Here only their effects relevant to our topic, namely spirituality, will be considered.

Three brain mechanisms—sedation, stimulation, and dissociation—are primarily associated with most drugs used in religious ceremonies. I do not mean to imply that these are the only ones; many others definitely exist. At the moment, they are not as well known. We will briefly consider these mechanisms and the drugs that produce them.

Sedation

Alcohol, which produces sedation, is most certainly one of the oldest and most ubiquitous drugs known to man. Around 6000 B.C., south of the Black Sea, the Neolithic people appear to have discovered fermentation of fruit juice and grains for the first time. Ancient Egyptians also knew of wines, and murals in pyramids depict wine making and consumption. They made a drink called *hak* from a grain. An alcoholic beverage made from grain was also popular among the Assyrians and the Babylonians. The Old Testament celebrated the joy and happiness alcohol brought about and supported its medicinal properties. However, excessive use was looked down upon.

Ancient Indians also knew about alcohol, and references to alcohol-containing beverages such as *sura* can be found in the Vedas. Intoxicating drinks were also distilled from molasses, fruits, grapes, the sweet sap of coconut palm tops, and so forth (Vishṇusmṛthi, XXII, 83). With a few exceptions, alcohol, however, had a minimal role in religious ceremonies.

Alcohol is an important part of the religious rites and rituals of Tantrism, an antinomian element in theistic Hinduism. Its adherents deliberately and consciously violate all taboos of Vedic Hinduism. The five M's *(makāras)* forbidden by orthodoxy—*māmsa* (flesh or meat), *matsya* (fish), *madya* (fermented grapes, wine), *mudrā* (fermented cereal and parched grain), and *maithuna* (sexual practices considered vulgar, incestuous, adulterous, and so on)—occupy a central role in their ceremonies. The underlying spiritual principle is the ultimate unity of the devil and the divine, the savage and the saint, the spiritual and the secular.

Union of the polarized opposites Yin and Yang results in the disappearance of differences and reemergence of the underlying unifying principle. When the negatively charged particle collides with the positively charged one, energy results. Tantrism revolves around union of the male (Śiva) and female (Śakti) elements; the female part is given special emphasis.

Alcohol, as part of the ritual, facilitates freeing consciousness from its contaminating worldly elements. Participation in these rituals without careful preparation and thorough understanding of the procedures is considered to be extremely dangerous. There are several subsects among the Tantras, and Tantric sects are present even within Buddhism and other Indian religions. Tantric Buddhism, otherwise known as Vajrāyaṇa, includes alcohol consumption in the quest for enlightenment.

Śakti, which is essentially goddess worship, is not synonymous with Tantrism. Tantrism is just one form of goddess worship. The goddess manifests in two forms, a motherly, benevolent, nurturing, and life-giving form and a malevolent, erotic, ferocious, and destructive form. As was mentioned above, both are derivatives of the same divine principle. Alcohol, meat, and sacrifices (including human) are used to propitiate and placate the malevolent form.

In ancient India, alcohol was used even by the orthodox. Early Vedic literature suggests the use of alcohol and animal products by the priestly classes. The Buddha is believed to have been responsible for blaspheming the ceremonial use of meat, fish, and alcohol. He allowed use of meat and fish, provided the creature was not killed specifically for that individual's consumption (Vinayapiṭaka, Mahāvagga, VI, 31, 14). He definitely enjoined cruelty to animals.

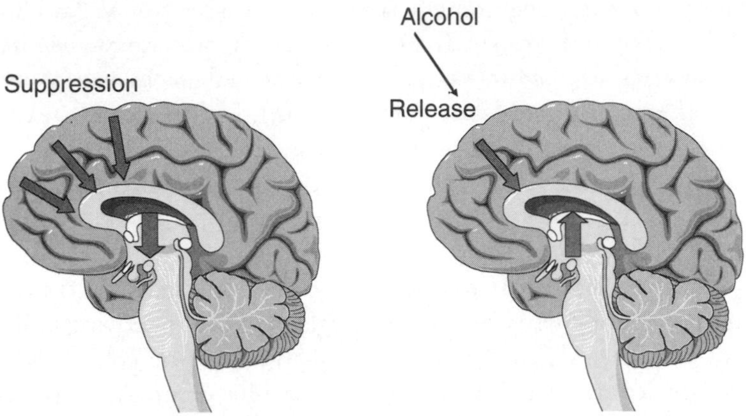

FIGURE 6.1 *Higher brain centers, especially the frontal cortex, inhibit lower brain centers. Drugs, by inhibiting the higher brain centers, release the lower ones from suppression, in a phenomenon called disinhibition.*

He forbade intoxication with both alcohol and drugs (Vinayapiṭaka, Cullavagga, XII, 1, 3).

Alcohol is a central nervous system depressant. It depresses evolutionally higher brain centers that normally inhibit lower ones. The inhibition of the higher centers leads to the release of the lower ones from inhibitory control; the phenomenon is called disinhibition (Figure 6.1). With small doses, the finer, more refined and sophisticated "higher brain functions"—logic, reason, social appropriateness, and the like—are weakened. Individuals become loquacious, euphoric, loud, and boisterous, and their behavior can become socially inappropriate. Normal inhibitions considered essential for civilized interactions become weaker. Individuals are known to become sexually inappropriate and exercise poor judgment. They can be provoked to violence more easily. With higher doses, they become nauseous and may vomit, their gait becomes unsteady, and their speech becomes slurred. At still higher doses, they become extremely sleepy, confused, and disoriented. With very high alcohol levels, they can go into a coma and eventually die. The wave of depressed activity seems to start from the most recent to the oldest brain structure and function, on a scale of evolution. Behavioral disinhibition is largely due to cortical suppression. As the blood alcohol level climbs, the midbrain and, eventually, the brain stem, which controls consciousness, are inhibited. Suppression of the brain stem causes death.

The frontal lobe, the crown of evolution, governs human behavior. While the role of the frontal lobe in suppressing the subcortical brain structures is well established, the exact subcortical structure or structures kept under frontal control is rather unclear. However, it is clear that alcohol-induced release produces happiness. Nobody drinks to become more socially appropriate and to gain better control; it is always for the opposite effects. The frontal lobe, which exerts inhibitory control and mediates reason, is the most recent addition to the brain, in an evolutionary sense. Lesions of the frontal lobes, including tumors, strokes, and degeneration, can be accompanied by disinhibition and euphoria, similar to that induced by alcohol.

That humans all over the world seek to inactivate the most sophisticated apparatus of brain evolution is of interest. That such inactivation should produce euphoria is even more interesting.

As we have seen, *ānanda,* or bliss, is intimately aligned with consciousness. Recently developed components of the mind suppress and distort this fundamental feeling of happiness. Inactivation of these appurtenances will lead to the free and unrestrained expression of the underlying bliss. Dreamless sleep, where higher centers of the brain are inactive, is also associated with bliss. Thus, it would seem probable that consciousness and bliss are closely related.

Neuroscience has not recognized such an association between consciousness and bliss, and consequently, very little is known about its physiology. However, some speculations are possible. Arousal mediated by the brain stem reticular activating system forms the physiologic substrate for consciousness.

As discussed in Chapter 2, neurotransmitters are chemicals that mediate communication between neurons. The reticular activating system is made up of several groupings of neurons and uses different neurotransmitters, of which norepinephrine is one. A cluster of norepinephrine-containing neurons in the brain stem, called the locus ceruleus, sprouts numerous tendrils containing the chemical. These nerve fibers traverse and innervate the brain at all levels. Locus ceruleus activation is central to arousal and consciousness. If there is an association between consciousness and bliss, it is likely to involve norepinephrine.

There are two risky hypotheses here: that there is an association between arousal and happiness and that there is an association between arousal-related happiness and norepinephrine. There is not a great deal of scientific evidence in support of either. However, there is not a whole lot of evidence against these possibilities, either.

Norepinephrine is listed among pleasure chemicals, but dopamine, a precursor of norepinephrine, is the archangel of pleasure. The dopamine-mediated pleasure pathways have been hypothesized to be evolutionary modifications of the phylogenically older norepinephrine-mediated arousal system. Norepinephrine mediates arousal, and it may also be related to pleasure. Thus, it would seem that living in its own right is a pleasurable experience. Even the most primitive creature values its existence, which is a clue that it must be associated with some happiness or pleasure.

Stimulation

Stimulation, the polar opposite of sedation, is the second mechanism involved in the central nervous system effects of drugs used to enhance spiritual states. Cocaine is the best example.

Humans started using cocaine at least 5,000 years ago. Coca leaves were found in burial sites in Peru dating from 2500 B.C. Colombian stone idols with puffed cheeks showed that cocaine chewing dated back to 500 B.C. Cocaine is present in the leaves of two species of coca shrub, *Erythroxylum coca*. This plant appears to be native to the Peruvian Andes, and now is grown in the Andes in Ecuador, Peru, and Bolivia. The drug is intertwined with fables and folklore of South American tribes. According to Colombian

Indians, their people came from the Milky Way in a canoe that contained coca and other drugs and was pulled by an anaconda.

The Incas acquired the drug when they conquered the region around the tenth century. They considered the plant of divine origin and its use a privilege restricted to priests and the upper classes. A lovely woman is depicted as mama coca, the spirit of the cocoa shrub. Coca formed an important component of their sacraments, and the Incas strictly controlled the cultivation and harvesting of the plant.

The Spanish brought cocaine and its use to Europe. A number of beverages, including John S. Pemberton's Coca-Cola in 1886 and Angelo Mariani's Cocoa Wine in 1883, became popular. Prominent individuals, including Sir Arthur Conan Doyle, Robert Louis Stevenson, Jules Verne, Thomas Edison, and Emile Zola, were cocaine users. Pope Leo VIII was so favorably impressed with Mariani's Cocoa Wine that the creator of the product was given a Gold Medal and the pope kept a flask of the wine with him at all times. Albert Niemann of Göttingen in Germany identified and isolated the active ingredient cocaine from the plant.

Concentration of cocaine in the plant is usually below 1.8 percent, and it was often chewed, mixed with lime, to increase absorption. The stimulation, euphoria, and addictiveness were, therefore, relatively low. Identification of the active ingredient and its extraction made administration of large quantities of the drug possible. While this increased the popularity of the drug, it also increased its adverse effects and addiction potential.

Cocaine stimulates the brain, and cocaine intoxication is associated with euphoria, expansiveness, talkativeness, restlessness, agitation, impaired judgment, grandiosity, impulsiveness, hypervigilance, and impaired judgment. It abolishes the need for sleep, and its subjects report feeling wired up.

Cocaine enhances arousal, which as we have seen is the physiological basis for phenomenal (empirical) consciousness. Activation of arousal will lead to enhancement of all activities it supports. Most, if not all, mental activities are energized and accelerated, and this forms the hallmark of cocaine intoxication. The predominant mood is one of euphoria. Autonomic nervous system stimulation results in rapid pulse and respiration and elevated blood pressure. The unnatural brain activation often results in impaired performance in a number of areas, including attention, concentration, and memory. Cocaine stimulates both norepinephrine and dopamine, but the euphoria is believed to be mediated primarily through dopamine stimulation.

Dopamine is not a fundamental component of arousal. However, as was discussed above, dopamine-related pleasure mechanisms may have evolved from norepinephrine-mediated arousal. Stimulants with a minimal effect of dopamine usually do not produce euphoria. For example, adrenaline is a stimulant, but it does not produce the mood enhancement cocaine does. But people do get addicted to such stimulants as nasal decongestants, caffeine, and diet pills.

Whereas sedatives release consciousness from inhibition, stimulants activate it directly. Both sedatives and stimulants have far-flung effects on brain and behavior. Their consciousness-enhancing properties are at best partial and contaminated. In addition, they are rarely, if ever, used in religious ceremonies at present. While they do provide some important cues, they do not tell us much more about the brain mechanisms related to spirituality. However, the story is very different with dissociative drugs. They have clear associations with religious ceremonies, both past and present.

Dissociation

According to the American Psychiatric Association's *Diagnostic and Statistical Manual* (fourth edition), dissociation is a disruption in the usually integrated functions of consciousness, memory, identity, and perception of the environment. This definition rests on the assumption that consciousness is dependent on the other functions of the mind and not the other way around.

Unfortunately, the manual provides no definition or description for consciousness. In the Upaniṣadic model of the mind, dissociation would entail a partial separation between the central core of consciousness and its veils, or *kośas*. According to the model, the *kośas* are appendages, and their association with consciousness is coincidental and not essential. Under normal circumstances, consciousness and the *kośas* are interlocked and experienced as a single whole. During dissociation, consciousness is detached from one or more of its accompaniments. Such a change will result in an altered perception of self. As noted previously, consciousness, the subjective feeling of life, is the quintessential subject. The subject is dissociated from the body in such dissociative states as hypnotic trances, when pain is not felt. In a condition called dissociative fugue, the subject loses contact with memory of person, place, and time, and the individual tends to wander away from home or the customary place of daily activities. In dissociative amnesia, the subject experiences an inability to recall important personal information, usually of a traumatic or stressful nature, that is too extensive to be explained by normal forgetfulness. It is associated with war experiences, rape, trauma, accidents, and so forth.

The most common type of dissociation is depersonalization, where the subject experiences a feeling of detachment or estrangement from the self. The individual may feel like an automaton or as if he or she is living in a dream or a movie. There may be a sensation of being an outside observer of one's mental processes, one's body, or parts of one's body. Various types of sensory anesthesia, lack of affective response, and a sensation of lacking control over one's actions, including speech, are often present. This condition has been found to occur in 40–50 percent of normal individuals, and it can be associated with a wide variety of conditions.

Dissociation is associated with loosening the bonds between consciousness and its appendages. Since consciousness is closely aligned with bliss, the experience of unfettered consciousness must be pleasurable. Release of consciousness through alcohol-induced disinhibition and activation by cocaine are both pleasurable. Yet depersonalization is considered to be an unpleasant experience.

There are two main reasons. First, a number of conditions associated with depersonalization are unpleasant in their own right. This would include fatigue, sleep deprivation, anxiety, depression, schizophrenia, temporal-lobe epilepsy, and temporal-lobe migraine. In many such conditions, depersonalization is triggered by the unpleasantness of the causal condition. In panic anxiety and severe depression, the mental pain becomes so intense that the consciousness (subject) detaches itself from the pain-containing mental apparatus *(kośa),* which may be the mind or the body. Depersonalization is triggered when physical pain, panic, and depression peak.

This was clearly documented by the famous British psychiatrist Malcolm Lader during an experiment on anxiety on a twenty-five-year-old woman. When she reported feeling like a robot and devoid of emotions, her anxiety measured by palmar sweat gland activity showed low levels. A few weeks later, when she reverted to her usual anxious state, the anxiety levels and sweat gland activity showed high readings. This finding, Dr. Lader argued, was consistent with the hypothesis that depersonalization was linked to some physiological mechanism for counteracting excess anxiety. For this reason, a number of experts have characterized depersonalization as a defense against unpleasant mood states.

As we have seen, other dissociative states such as fugue and amnesia are also triggered by intense mental or physical pain. In these conditions, depersonalization is not the cause of pain. In fact, it provides relief from the pain caused by the primary factor. Depersonalization in other conditions, such as confusional states, schizophrenia, and epilepsy, does not involve pain or relief from it. It is conceivable that in these disorders, the neural mechanisms that couple consciousness to other mental functions are seriously disabled. This, in turn, may lead to the weakening of the links between consciousness and the rest of the mind as well as depersonalization.

Primary depersonalization disorder is a condition that exists on its own, with nothing causing it. Individuals with this disorder who see psychiatrists do so because they find the condition unpleasant. Depersonalization causes a dramatic and drastic mental state change, which can be frightening. I have seen patients with severe depression and anxiety petrified by their depersonalization experience, although it brought relief. A number of them had a shuddering fear of losing their minds. In these cases, the anxiety is not caused by depersonalization per se but rather by the individual's reaction to it. Many, however, do not find the experience unpleasant and therefore do not seek professional help. For obvious reasons, they are underrepresented in psychiatric literature.

The situation is completely different for individuals who purposefully ingest drugs or engage in such practices as meditation and yoga to achieve the effect. These individuals look forward to depersonalization and consider it quite blissful and pleasurable.

In prehistoric days, drugs and practices that induced dissociation seem to have played an important role in religious ceremonies. Primitive man, unbound by social mores, external appearances, and social propriety, was more willing and able to let go and immerse himself in the stream of experiences these drugs and practices unleashed. These practices can still be found in isolated pockets in the modernized world and in parts of the world less touched by intellectualism, materialism, and prudery. People from developed countries who allow themselves to partake in these ceremonies are often astounded at the scope and sweep of these experiences. Visions hitherto unknown are displayed, uncharted depths stirred, profound insights gained, and firmly established concepts of reality and normality shattered. Away from civilization, in a less-sophisticated culture where life is simple and natural, one gets a glimpse of the shallowness of one's convictions of self and the experience of self. Unfettered by social bounds and barriers, one can feel true freedom to explore and experiment, without fear and apprehension.

Dissociative drugs are otherwise known as hallucinogens. They bring about changes in perception, thought, and mood, but they seldom produce mental confusion, memory loss, or disorientation for person, place, and time. They do not produce quantitative changes in consciousness either toward sedation or stimulation, but they do alter the quality of the conscious experience. Most, but not all, of the hallucinogenic compounds are vegetal. We know about approximately 150 of them, but experts estimate that about 500,000 species exist. In spite of its protean fauna and flora, the African continent is relatively sparse in plants of this type. One such substance, iboga, is used in Gabon and parts of the Congo in the Bwiti cult. In Botswana, the natives rub slices of the bulb of kwashi over their heads in sacramental ceremonies. In Asia, fly agaric, known as soma in ancient India, was popular. Datura also enjoyed widespread use in Asia. Other drugs used there include nutmeg, aracholine, and marijuana.

Most hallucinogens used in medieval Europe belong to the *Solanaceae* family. Thornapple, mandrake, henbane, and belladonna fall within this grouping. Accidental ingestion of ergot-contaminated rye caused St. Anthony's fire, and in addition, insanity, hallucinations, gangrene, even death. To the best of our knowledge, however, this drug was not used in any religious ceremony.

Similarly, hallucinogenic drugs were not abundant in the American continent north of Mexico. Psilocybin is the active ingredient in several species of mushrooms native to the American subcontinent. Central and South American countries, and especially Mexico, undoubtedly provide the richest supply of hallucinogenic plants. Peyote cactus is the single most important sacred hallucinogen there. Twenty-four species of mushrooms have been found to contain hallucinogenic compounds, and many other hallucinogens have also been found in this region. In the Amazon basin, a drink called *ayahuasca caapi,* produced from the stem bark of a vine, was used to produce a number of supersensory effects, including clairvoyance. It is alleged to have produced a sensation of flying.

Of the wide variety of drugs that have been used in religious ceremonies I would like to discuss three better-known and researched ones: peyote, LSD, and marijuana.

Peyote

The peyote cactus is a small, stumpy, gray-green or bluish-green plant with a thick, tapering root. The roughly hemispherical crown that protrudes above the ground is broken up into five to thirteen rounded ribs by shallow furrows radiating from a small, circular, hairy center (Figure 6.2). The cactus with the botanical name *Lophophora williamsii* contains around thirty alkaloids,

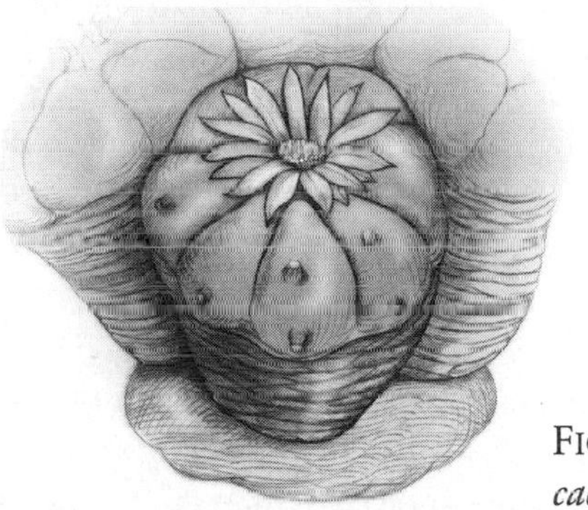

FIGURE 6.2 *The peyote cactus containing mescaline.*

including mescaline. *L. diffusa*, the San Pedro cactus *(Trichocereus pachanoi)*, and several others, have been found to contain the same chemical.

Mescaline has a long and tortuous association with religious ceremonies. The Spanish conquistadors, the first Westerners to come across this unusual drug, found ceremonial use of the drug firmly established. The Spanish chroniclers estimated its use among the Chichimeca and Toltecs went back 2,000 years or more. Rock carvings of the peyote ceremony and archeological discoveries in dry caves and rock shelters in Texas, however, clearly indicate that its use is more than 3,000 years old. Some investigators believe that the drug was first discovered by Tarahumara Indians and spread from them to other tribes.

The peyote cactus had both medicinal and religious uses, and it titillated sufficient interest in Europe for King Philip II of Spain to send Dr. Francisco Hernandez, his personal physician, to the New World to investigate the matter further. Dr. Hernandez did not arrive at any firm conclusions except to state, “It causes those devouring it to be able to foresee and to predict things.”[1]

Predictably, the Spanish missionaries were not impressed; they regarded it as diabolical. They also noted that the drug was combined with wine and that the ceremonies involved singing and dancing. These peyote ceremonies were always group events.

Catholic missionaries eager to convert the natives living "in sin and infamy" saw no place for peyote in the religion they were attempting to promulgate. Peyote use was equated with cannibalism and vampirism, and peyote users were persecuted.

Peyote ceremonies persisted in spite of the forced oppression, and they continue to date. A peyote ceremony may be held at any time for any number of reasons, including health, wealth, or worship. The ceremony itself is a group event and is associated with praying, dancing, singing, and so on. These ceremonies, which begin after sundown, usually last the entire night.

Peyote is made divine as Tetewari among Huichols. Under the guidance of the Shaman, present-day Huichol pilgrims go in search of peyote cactus in a highly ritualized ceremony. In olden days, the Indians walked hundreds of miles; the trip at present is made by bus and automobile. Ritual confession and purification through abstinence are essential, with negative thoughts, feelings, and desires expunged, and carnal impulses assiduously avoided. The peyote "hunt" is punctuated with chanting and praying. The entire event is steeped in symbolism and mythology. The collected peyote is taken home and also sold to other native tribes who do not conduct a quest of their own.

North American native tribes enthusiastically adapted peyote use in their own religious ceremonies. The Kiowa and Comanche were the first to acquire the practice from the Mexicans. Native American tribes, with their age-old customs and practices decimated, took to the peyote ceremonies readily. The peyote religion swept across the plains and, later, to other tribes. Under the influence of the ecclesiasts, local governments enacted laws that criminalized its use.

The Native American Church integrated peyote into its Christian ceremonies. The natives, unlike the Europeans who brought Christianity to them, perceived no incompatibility between Christianity and peyote use. They were able to blend the two effortlessly. However, the practice was declared illegal on a number of occasions. Finally, in 1960, Judge Yale McFate of Arizona, in a roundhouse ruling, overturned previous federal and state laws banning Native Americans from using the drug and granted Fourteenth and First Amendment protection, on the grounds of religious freedom.

Peyote ceremonies highlight a variety of religious and social events, including birthdays, baptisms, funerals, memorial services, even Easter, Thanksgiving, Christmas, and the New Year. Both men and women participate, but children are not allowed to consume the drug. Drumming, singing, and dancing enter into the celebration, and peyote consumption is regarded as a sacrament. Peyote, according to many, symbolizes the power of Jesus.

Scientific study of peyote probably started in the latter part of the 1800s when Parke-Davis and other pharmaceutical enterprises started showing some interest in it. The Philadelphia physician Weir Mitchell and Havelock Ellis, a research psychologist, provided the first scientific description of inebriation with peyote. In 1897, Arthur Heffter identified the chemical responsible for most of its effects, dubbing it mescaline. Since then, there have been a number of descriptions of mescaline intoxication, and isolated groups of individuals in North America and Europe began to ingest the drug for its epiphany-producing properties.

Pharmacological research on mescaline and its effects caught the eye of the famous novelist Aldous Huxley. Huxley ingested mescaline sulfate in 1953 and repeated the experiment several times. In 1954, he reported his experiences in *The Doors of Perception*. The book sparked a fierce controversy, with well-known people lining up on either side. Although neither Huxley nor any of the peyote proponents started a peyote cult in the Western world as such, the book definitely introduced a number of people to the drug. At present, peyote is a Schedule I drug (the Drug Enforcement Administration classification) with no accepted medicinal use.

The drug is usually ingested, although mescaline hydrochloride may be administered by injection. Different ingenious ways have been devised to mask its bitter, unpalatable taste. Peyote enthusiasts dismiss the gastric discomfort caused by the drug as trivial in comparison to the profound mental experience of peyote inebriation. In fact, some peyote users argue that there is a direct relationship between the severity of gagging and the profundity of the mystical experience.

The initial mental changes make their appearance approximately one hour after the oral consumption of the drug. Intoxication intensifies over the next three to five hours and subsides in another five to six hours. The duration is considerably lengthened by repeated consumption of the drug, as is typical in native ceremonies.

The drug effect is highly variable. Descriptions of the mental effects of the drug are restricted by the limitations of the medium of language. A great deal of what is experienced cannot be contained in the concrete, linguistic elements. Users frequently describe the experience as profound, insightful, divine, beatific, and mystical. But what exactly these terms mean is vague and unclear.

First of all, dissociation, especially depersonalization, is extremely common. Perception is altered at all levels, from basic to most refined. Illusions and hallucinatory experiences are quite common, and both formed and unformed sensory images are perceived. Eidetic and spectral images of protean shades and colors appear, disappear, and melt into one another. The colors take on an intensely pleasing, unearthly hue and glow.

Hallucinations of familiar and unfamiliar objects—heaps of glittering jewels, smoke-puffing dragons, winged angels—have all been reported. Hallucinations may occur in all sensory modalities—visual, auditory, and somesthenic. The foundations upon which perceptual reality is built are violently shaken, and rules of normality are outright violated. Subject-object distinctions are toned down or completely obliterated. The perceiver becomes part of the perceived. Lines of cleavage between sensory modalities soften, and music is seen and fragrances heard. The phenomenon of overlapping of different senses is called synaesthesiae. Body perception is qualitatively and quantitatively altered. The body shrinks down to a midget in relation to the surroundings or enlarges into a giant and may become luminous or transparent.

Time and space are distorted and altered. Time usually slows down and may even come to a screeching halt. Flow of time may become irregular, moving in fits and starts. The fundamental sense of a steady flow of time is almost always completely gone.

"Out-of-body" experiences are frequent accompaniments of peyote inebriation. This is associated with the sensations of floating and flying. There is a sense of detachment, peace, and freedom, and all negative emotions are effaced. A sensation of buoyancy and expansion are often present. For most people, spiritual and religious themes become the center of their thinking process, with these religious thoughts and feelings dictating and directing their mental operations.

Native American Church members' experiences are strongly colored by Christianity. In one report, Jesus appeared to a woman in bereavement and consoled her about her son's untimely demise.

A reputed Western thinker of Buddhist persuasion experienced the "three great truths"—Buddham (enlightenment), Dharmam (the cosmic code of ethics), and Samgham (the community of monks)—he had long accepted intellectually. He experienced an awareness of undifferentiated unity, bliss, and all that is implied by the Buddhist doctrine of Dharmam. Atheists have also been reported transforming into theists under the influence of peyote.

Creativity and aesthetic appreciation are enhanced, and the eidetic images seen during peyote experiences find expression in tribal art forms. While a certain degree of confusion is inevitable, overt confusional states and total psychotic breakdowns are rare, but they have been reported.

Studies have also been conducted on the effects of mescaline on the brain. The German scientists Leo Hermle and associates examined the acute effects of mescaline on cerebral blood flow in twelve normal subjects. Mescaline increased flow to the frontal lobes, with special emphasis on the nondominant right hemisphere. In another study of six healthy male volunteers, the effect of mescaline on balance between the two hemispheres was studied. Hemispheric effects were evaluated with a psychological test and cerebral blood flow measurement. The findings suggested excitation of the right hemispheric activity as mescaline intoxication increased in intensity.

LSD

The Aztecs used a plant they called Ololiuqui (Morning glory, or *Turbina corymbosa*) in their religious sacraments and sacrifices. This plant contained a potent hallucinogen, lysergic acid amide. Lysergic acid diethylamide (LSD) was a product of pharmacological research, and its use in religious ceremonies is contemporary.

The second of the three drugs that have been used in religious ceremonies, LSD, traces back to Dr. Albert Hofmann of Sandoz, a pharmaceutical company based in Basel, Switzerland, who accidentally found the drug while synthesizing various ergot analogs. The drug was expected to have a stimulating effect on circulation and respiration, and in animal experiments, it caused uterine contractions. Although the drug was initially written off, Hofmann had a sudden intuitive insight concerning the drug's yet unknown, but useful, effects. In the spring of 1943, he decided to take a second look at the drug. While recrystallizing a very small quantity of LSD, he had a hallucinogenic experience; he had accidentally absorbed a small quantity through his skin. The intoxication lasted less than two hours. Convinced the drug had central nervous system effects, he ingested a very small quantity (250 micrograms) without realizing that the compound he had found was approximately 4,000 times as potent as mescaline.

Forty minutes after ingesting the drug, he experienced its psychedelic effect loud and clear. He became so impaired that he could not write. He and a laboratory assistant bicycled home. Dr. Hofmann wrote: "While we were cycling home, however, it became clear that the symptoms were much stronger than the first time. I had great difficulty in speaking coherently, my field of vision swayed before me, and objects appeared distorted like the images in curved mirrors." Concerned he was losing his mind, he sent for a doctor. Six hours later, his mental status began to come back to normal. Other people at Sandoz repeated the experiment with one-third the dose and had comparable effects.

Born in a prestigious pharmacological laboratory and developed under the tutelary guardianship of scientists, LSD was destined to become a psychopharmacological agent for use in research and treatment. Hofmann's descriptions of out-of-body experiences, transcendence, detachment of ego from the body, and "deeply religious" feelings did not receive much attention at Sandoz. They saw the drug primarily as a key to the mysteries of brains afflicted with schizophrenia. The drug was exposed to intensive research, wide in scope and variety. Many descriptions of the LSD-induced state of mind can be found. Although researchers found very obvious similarities between mescaline and LSD, the dysphoric effects and silliness such as giggling, laughter, and experiences of somatic pain appeared much more common with LSD.

One has to bear in mind that whereas peyote ingestion was a religious ritual for truly devout people seeking communion with God, most of the LSD experiments were done on volunteers in laboratories.

LSD did not have much use in schizophrenia research, as there were glaring differences between LSD intoxication and the psychotic disorder. However, a number of psychiatrists and psychologists became interested in the use of LSD as an adjunct to psychotherapy. Unlike peyote, LSD was a proper pharmacological agent, studied, with the use of rigorous research methodology. The effective dose was firmly established, and no serious adverse effects were proven, although several were alleged. A number of physicians, psychologists, philosophers, anthropologists, socialites, and prominent citizens tried LSD. While some found the effects distinctly unpleasant, most were awed and overwhelmed.

The mystical effects of the drug profoundly impressed and influenced two Harvard psychologists, Timothy Leary and Richard Alpert. They were willing to sacrifice their Harvard careers in pursuit and popularization of LSD use. Leary and his associates established a number of foundations (the International Federation for Internal Freedom, the Castalia Foundation) and publications *(Psychedelic Review)*, and they traveled far and wide extolling the virtues of the drug and propounding the spiritual significance of the experience and its benefit to human existence. LSD use became so widespread that in 1967, the alarmed authorities established a federal ban on the drug.

When Alpert and Leary were forced to leave Harvard, the pendulum began to swing to the other side. The adverse effects of LSD, including psychosis, personality change, and criminality, were highlighted. The social opprobrium strengthened, and the illegality of LSD became firmly rooted. In 1977, the National Institute on Drug Abuse concluded that 6 percent of the population above the age of twelve (10 million people) had used a strong psychedelic, mostly LSD, with 1 million being regular users.

Both LSD and mescaline have been found to produce a variety of changes in the EEG, but most suggested increased arousal, evidenced by decreased amplitude and increased frequency of brain waves. EEG experiments, which compared its effects on the two hemispheres, concluded that LSD preferentially stimulated the visuospatial right hemisphere that under normal conditions is nondominant. Under normal conditions, the analytical left hemisphere, which mediates logic and reason, supersedes the other hemisphere. The nondominant hemisphere uses a holistic modus operandi, and it is believed to mediate appreciation of art and music.

The mescaline ceremony was associated with singing, chanting, drumming, and dancing. LSD, the hallucinogen of the "civilized world," also found a disproportionate number of its supporters among artists. A number of artists and musicians took to LSD wholeheartedly, and a special type of art, called psychedelic art, emerged. A number of prominent musical groups overtly and covertly supported its use. An independent panel of critics found paintings by prominent artists made under LSD to be of greater value.

Accredited researchers William McGlothlin and Sydney Cohen published research findings that supported an association between creativity and aesthetic appreciation and LSD intoxication in prestigious scientific journals such as the *Journal of Nervous and Mental Diseases* and the *Archives of General Psychiatry*. After LSD sessions, more of their research participants reported a greater appreciation of music. More purchased records and spent more time in musical events and museums. Cohen argued that LSD enhanced the subjective feeling of creativity. However, others found that giving the drug to an unselected group of people did not significantly enhance their creative ability. A number of artists surveyed reported beneficial effects of LSD on their artistic abilities. Thus, it would appear that psychedelics and art have a mutually supportive and reinforcing effect.

Houston Smith, professor of philosophy at Massachusetts Institute of Technology, as well as others, wrote about the religious feelings induced by LSD. They argued that drugs could induce religious experiences indistinguishable from ones that occur spontaneously. In a survey Smith conducted with James Masters, 96 percent of the individuals surveyed reported seeing religious imagery of some kind. Fifty-eight percent saw religious personages such as Christ and the Buddha. Large numbers visualized religious architecture, symbols, and shibboleths.

As might be expected, as restrictions and regulations on LSD and mescaline increased, interest in research involving these compounds also declined, thus very little research with the more recent sophisticated brain-imaging techniques has been conducted with these two compounds.

Marijuana

The third of the three drugs used in religious ceremonies, marijuana, *has* been studied. Today, the single most popular dissociative drug in the world is marijuana. Its use is as ancient as alcohol, and most ancient civilizations seem to have known the drug. Findings suggestive of earliest use come from the island of Taiwan over 10,000 years ago. The Chinese word for cannabis is *ma*. Emperor Fu Hsi (2900 B.C.), during whose reign Yin and Yang concepts were developed, noted that *ma* was endowed with both Yin and Yang.

Marijuana, a member of the hemp family, was useful to the ancients in a number of ways, including medicinal use and fiber for fabric. The Chinese seem to have been the first people to note the intoxicating effects of the drug. Taoists, who permeated China, knew about marijuana intoxication but condemned it as the liberator of sin. During the early part of the Common Era, the Chinese were using cannabis seeds in their incense burners to induce intoxicating effects. Dating back around 600 B.C., the Zend-Avesta of ancient Persia refers to an intoxicating resin, presumably marijuana.

The Assyrians used cannabis as incense in the ninth century B.C. In ancient Greece, Galen reported the use of marijuana at parties. In the African subcontinent, it was used by many native cultures in social and religious situations. Known as kif, or *duagga*, its vapors were inhaled by devotees from burning hemp. The Kasai tribes of the Congo smoked marijuana from calabash pipes. The Tepecano Indians of Northwest Mexico are reported to have substituted for peyote with cannabis when the former was not available. Mexican natives still use cannabis under the name Santa Rosa in a Christian ceremony involving the Virgin.

Scythians, a warring nomadic Middle Eastern tribe, are supposed to have christened the plant cannabis and to have been responsible for spreading its use. The Scythians were closely related to the Semites, and there has been a great deal of controversy as to whether the ancient Jews knew of cannabis. The holy oil God instructed Moses to make from "myrrh, sweet cinnamon, *kaneh bosn*, and Kassia" is believed to contain cannabis. Previous translations of *kaneh bosn* as "calamus," a marsh plant, were found to be erroneous; the Hebrew *kaneh bosn* and Scythian cannabis were probably the same. The Scythian obsequies, including cleansing by burning cannabis seeds, are still practiced. Scythians took cannabis into Egypt, via Palestine, and north into Russia and Europe.

Islam firmly banned all intoxicating agents, including marijuana. However, the Sufis, an unorthodox offshoot of Islam that considered spiritual experiences and communication with the Almighty central to their belief, used hashish on a large scale. Orthodox Muslims do not look upon Sufism favorably, however, and its followers are often on the fringes of Islam. Hasan-sabbah, a Muslim extremist who attempted to cleanse Islam through the method of secret assassination, is believed to have rewarded his murderous followers with a drug that opened the paradise gates. Although this divine drug probably was opium, early storytellers thought it was hashish. In fact, the name "hashish" was derived from Hashishin, Hasan's full name. His name is also associated with the term "assassin."

Marijuana, otherwise known as Indian hemp, has a close and intimate association with the Indian subcontinent. The earliest Indian text that refers to the drug is the fourth Veda, Atharva Veda. Atharva Veda is believed to be the forerunner of Ayurveda, the Indian system of medicine. Many believe that Atharva Veda was derived largely from the pre-Āryan Indus Valley civilization, which would indicate that marijuana was known to the inhabitants of the Indus Valley before the arrival of the Āryans around 1500 B.C. In spite of the Vedic references to the drug, there is no evidence that the drug was used as part of the Vedic religion. Although never a component of orthodox Vedic religion, it has been part of the unorthodox, antinomian revolutionary movements within the far-flung mosaic of Hinduism.

Ancient Indians associated at least two plants with their religious ceremonies, soma (mentioned earlier) and cannabis. *Amanita muscaria,* the mushroom from which soma in extracted, grows in forests of pine, spruce, or birch. While these trees are abundant in the slopes of the Himalayas and further north, they are sparse in the Indo-Gangetic plains and the South. On the other hand, cannabis grows in abundance throughout India, especially in the northern regions. Atharva Veda, which probably was derived from the people who inhabited the pre-Āryan Indus Valley, mentions cannabis. Therefore, it is likely that cannabis was a hallucinogen that the Indus Valley people used. The victorious Āryans preferred their drug, soma. Cannabis use has apparently never been a part of Vedic Hinduism, although the unorthodox sects espoused its use. The Hindu God Śīva, whom some believe to be a later version of Pasupati, an Indus Valley deity, became associated with cannabis. In Ṛg Veda, he is called Rudra. He is known as the Lord of Bhang, a mild liquid extract made with marijuana leaves.

Āryans deified soma, the drug they brought, while the non- and anti-Āryan groups preferred the indigenous cannabis. Traditional Hinduism is a direct descendant of the Vedic religion of the Āryans. The Vedic religion was built around the caste system and

Āryan superiority that many found distasteful. Ṛg Veda refers to groups of apostates who were foreign to the mainstream religion. The longhaired ascetics (Keśins) and the reticent ones (Munis) receive special mention in the Veda. They existed on the outskirts of the society in defiance of accepted standards of normality. Many grew their hair and beard, and some went around naked. An aura of mystery and mystique surrounded them. They were viewed with admiration, fear, and curiosity. In many ways they are reminiscent of John the Baptist, who lived in the desert, donned animal skins, and existed on locusts and wild honey. Buddhism and Jainism were derived from such non-Vedic factions of Indian religion.

Ṛg Veda refers to a "poison" *(viṣa)* in association with the longhaired ascetics and the god Rudra. "The wind had churned it [the drug] up; Koonamnama [a hunchbacked female deity] prepared it for him. Long-hair drinks from the cup, sharing the poison with Rudra" (10, 136). There is no doubt that in later years Rudra transformed into Śīva, who is associated with Bhang. It is of interest that Śīva is excluded from soma ceremonies. According to Ṛg Veda, this drug enabled its users to see visions and leave their bodies and endowed them with superhuman powers. Rudra is also associated with datura, and whether the Ṛg Vedic "poison" is datura or cannabis is unclear. Very few people in contemporary India use datura, while large numbers of people, especially the heterodox sects, indulge in marijuana. However, the drug enjoys social acceptance and nonceremonial use, especially during festive occasions.

According to some documents, before his enlightenment, during the period of severe asceticism, the Buddha himself used to consume one cannabis seed a day. Although it has been claimed that this was meant to cleanse his body of sins, the exact purpose is unclear. It would seem highly unlikely that a single marijuana seed would confer any intoxication to speak of. Conservative Buddhism places tremendous importance on a simple, normal, and natural life free from passions and desires. In orthodox Buddhism, there is no room for any intoxicants, including cannabis, in it. However, Buddha did not object to his followers getting medical treatment, and he, himself, had a physician (Vinayapiṭaka, Mahāvagga, VIII,

1, 30). Since medicinal use of cannabis is mentioned in old Ayurvedic texts, it is possible that he permittd the use of the drug for medicinal purposes.

Tantrism, mentioned in relation to alcohol, also gives a prominent place to cannabis in rites and rituals. In rituals involving the goddess Kāli, Bhang is consumed to heighten senses to accomplish union with the goddess. Kāli depicts the malefic aspect of the goddess. Tantric Buddhism, which is popular in Tibet, also utilizes cannabis in its meditative rituals. Cannabis is a popular recreational drug in Tibet in general.

Cannabis contains a number of chemicals collectively referred to as cannabinoids, of which delta–9-tetrahydrocannabinol (THC) is considered to be the most important. THC is a resinous material with high solubility in fats and oils and very low solubility in water. The most common mode of administration is smoking, although the drug may be ingested as a beverage or baked into a cookie. The smoked drug is rapidly absorbed from the lungs, and inebriation occurs within a few minutes. The effects of the drug are usually gone in about three hours.

Marijuana smoking is followed by a wide variety of symptoms—euphoria, anxiety, lethargy, drowsiness, confusion, memory defects, altered time sense, depersonalization, impaired performance, cognitive changes, and psychosis. Negative symptoms, especially panic, are more common among inexperienced users or following intoxication with unusually large quantities of the drug. Detailed discussions of the positives and negatives of marijuana use, its benefits and evils, and its constructive and destructive effects on society do not concern us here. Rather, I would like to focus on three symptoms, which are of special relevance to us, namely intoxication, altered time sense, and depersonalization.

Over the years, several investigators, including the Duke University group, have examined the effects of marijuana on the brain and behavior, and a number of research reports are available on the subject. Our experiments at Duke involved administering marijuana, which is illegal in the United States, to human subjects and were conducted after obtaining approval from the Food and Drug Administration, Drug Enforcement Administration, North Carolina Center for Controlled Drugs, and Duke Institutional Review Board. The protocols were reviewed and approved, and the projects funded by the National Institute on Drug Abuse. Participants were marijuana users who were physically and mentally healthy and able to understand and give consent. The research protocols were described to the participants, and their written consent obtained. In some studies the participants smoked marijuana, while in others, THC (the active ingredient) was given as an intravenous infusion over twenty minutes. Different doses of the drug were used in different experiments.

The vast majority of subjects found the experience pleasurable. The exceptions were those who were anxiety prone or excessively anxious about the laboratory and the medical environment and individuals whose routine marijuana use was much lower than the amount they received as part of the experiment. In spite of this, however, of the large numbers of investigations conducted and reports published, there is no report that fully captures the subjective experience of marijuana inebriation. Since I have never smoked marijuana, I have no firsthand knowledge. I am simply reporting what was described to me by most of our subjects.

Almost all of them mentioned their inability to put into words exactly how they felt. Of the many rating scales used, none could faithfully quantitate the essence of marijuana intoxication. Although the effects may be ineffable, approximations are possible. It is not exciting like cocaine; it is more calming or tranquilizing. It does not energize; it induces a sense of pleasant lethargy. The pleasure is closer to listening to music or viewing a panorama, and not similar to winning a debate or sexual orgasm. Most considered it to be a very desirable state of mind. There very definitely is a sense of slowing of the passage of time. Usually time drags when one is bored; however, with marijuana the slower pace of time is accompanied by the calm repose of rest and relaxation. Both internal and external time slows down. The subjective feeling of the passage of time is affected, and clocks seem to tick slower, the needles of the watch to freeze.

Although we could not capture the qualitative aspects of the intoxication, we were able to measure its severity with a simple analog scale (a 10-cm line with one end representing no effect and the other the most intense). This enabled us to study the relationship between the subjective aspects of marijuana intoxication and other relevant physiological and psychological variables. We also studied the effects of marijuana on time sense.

In order to identify the brain region(s) associated with various behavioral phenomena related to marijuana intoxication, cerebral blood flow (CBF) was measured before, and several times after, the administration of marijuana or THC. For comparison purposes, similar measurements were made after placebo administration. In most experiments, the laboratory personnel and the participants did not know whether the subjects were smoking marijuana or the placebo. Earlier experiments utilized ^{133}Xenon Inhalation Technique for CBF measurement. More recent studies were performed with the technologically more sophisticated Positron Emission Tomography (PET).

Cerebral blood flow to both hemispheres showed significant increases after cannabis and THC, but not after the placebo. The increase was more pronounced in the nondominant right hemisphere. All participants were predominantly right-handed. Frontal regions showed more marked increases. Intoxication significantly correlated with the global increase in blood flow. Of the various brain regions studied, the frontal lobe and cingulate gyrus showed closest associations with levels of intoxication.

Conditions associated with high levels of arousal show an increase in blood flow while low arousal states show a decrease. The close relationship between marijuana intoxication and brain blood flow increase would, therefore, indicate that marijuana increases arousal. Arousal is the physiological substrate for phenomenal consciousness. The frontal lobe is the cortical representative of the arousal mechanism, and the selective increase in frontal activity following marijuana intoxication supports the notion that marijuana increases arousal levels. Like LSD and mescaline, marijuana also stimulated the nondominant right hemisphere selectively.

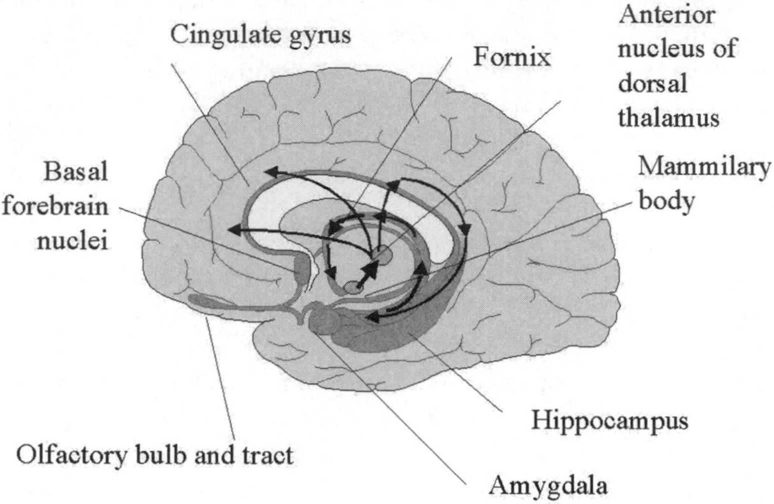

FIGURE 6.3 *The Papez circuit that controls emotions.*

Depersonalization, which is altered feeling of self (consciousness), was measured with a twelve-item questionnaire developed by J. C. Dixon in 1963. This scale was previously used to study marijuana intoxication by other investigators. Depersonalization set in and peaked thirty minutes after marijuana smoking and returned to the baseline in about 120 minutes. Marijuana of greater potency induced more severe depersonalization. However, all participants did not report this effect; about 30 percent definitely had the response and another 30 percent reported some depersonalization, while the others had none.

Depersonalization was most closely associated with blood flow to the right anterior cingulate. The anterior cingulate is an important component of the brain circuit—the Papez circuit—that mediates emotions (Figure 6.3). The anterior cingulate is the organ that renders conscious awareness to otherwise unconscious activities of the brain. According to Papez, the cingulate gyrus is the "seat of dynamic vigilance by which emotional experiences are endowed with an emotional consciousness."[2] This brain region, transposed between subcortical and cortical structures, may be responsible for integrating newer and older brain structures that contribute to the composite feeling of self.

Selective functional impairment of this region may result in the decoupling of these mechanisms, and dissociation may be the resultant behavioral manifestation. Electrical stimulation of the human cingulate has been reported to elicit a broad variety of behavioral changes, including altered states of consciousness. Some of the patients reported euphoria and a sense of well-being. Seizures with onset in the cingulate region also produce altered consciousness.

Drugs that alter consciousness usually produce altered time sense as well. In order to examine this possibility, our group at Duke measured time sense with a fourteen-item questionnaire developed by a friend and colleague of mine who is no more, Frederick Melges. Studies conducted with this "Temporal Disintegration Inventory" found altered time perception after marijuana smoking. Our research confirmed previous reports and, in addition, found an association between slowing of the passage of time and decreased blood flow to a brain structure called the cerebellum. The correlation approached statistical significance, but it did not quite make it, and therefore, this finding can best be described as provisional awaiting confirmation.

The relationship we found between decreased cerebellar activity and impaired time sense is in keeping with several previous reports that linked the cerebellum to an internal timing system. Other investigators showed that the cerebellum was involved in the temporal sequencing of motor activity. In human subjects, cerebellar lesions were associated with impairment in rhythmic tapping, a time-dependent task. Patients with cerebellar atrophy were found to be deficient at judging the relative duration of time intervals. Traditionally, the cerebellum has been associated with maintaining balance and coordination of movements. The present findings suggest that it is also responsible for the timing function essential both for balance and coordination.

Achieving Spirituality

Since antiquity, humans have used drugs that lift them above the liminality of the senses and the related mundane existence into a world of boundless freedom and joy. Sedatives like alcohol inhibit brain centers responsible for higher neural functions, especially those involving logic, reasoning, and ratiocination, thereby unleashing the previously bound consciousness. Stimulants such as cocaine expand and enlarge consciousness. Hallucinogenic drugs produce the effect through a number of mechanisms, of which depersonalization seems to be of special importance. The euphorias induced by sedatives, stimulants, and dissociative drugs are quite different.

These differences may be due to any number of factors, with the degree of activation of different components of the arousal chain being a serious possibility. The drugs activate mechanisms other than consciousness enhancement that may also account for the qualitative differences. Although sugar, for example, is sweet, coffee sweetened with sugar and Coca-Cola do not taste the same. The neurochemical basis for drug-induced euphoria is a topic of contemporary scientific research. Dopamine is generally accepted as the most important one. The relationships between disinhibition, activation, and dissociation and dopamine-mediated neurotransmission have received very little scientific attention.

According to Richard Evans Schultes and Albert Hoffmann (the latter was the first to synthesize LSD):

> *In general, we experience life from a rather limited point of view. This is the so-called normal state. However, through hallucinogens the perception of reality can be strongly changed and expanded. These different aspects, or levels, of one and the same reality are not mutually exclusive. They form an all-encompassing,*

> *timeless, transcendental reality . . . In normal states of consciousness—in everyday reality—ego and outside world are separated; one stands face-to-face with the outside world; it has become an object. Under the influence of hallucinogens, the borderline between the experiencing ego and the outside world disappears or becomes blurred, depending on the degree of inebriation. . . . These states of cosmic consciousness which under favorable circumstances may be attained with hallucinogens, is related to the spontaneous religious ecstasy of the unio mystica or, in the experience of Eastern religious experience as samadhi or satori. In both of these states, a reality is experienced which is illuminated by that transcendental reality in which creation and ego, sender and receiver, are One.*[3]

I am by no means suggesting, recommending, or even condoning the use of drugs to induce spiritual experiences. In fact, I would strongly argue against it.

Spirituality is the totality of the means and the goal. How one gets spiritual is more important than getting spiritual. Taking a pill to produce a paranormal state of mind and reaching a comparable state of mind through character improvement, austere practices, and devotion are very different.

According to Vivekananda, the best known of Indian philosophers of living memory (1863–1902), "The yogi says there is a great danger in stumbling upon this [spiritual] state. In a good many cases, there is the danger of their brains being deranged and as a rule, you will find that all these men, however great they were, who stumbled upon this superconscious state without understanding it, groping in the dark, and generally had, along with their knowledge, some quaint superstition."[4] Timothy Leary exemplifies Vivekananda's point vividly.

Whether these drugs do indeed produce states of mind identical to those associated with spiritual practices is difficult to verify, both being highly subjective and ineffable.

Even if we grant that the drugs produce spiritual states of mind, drug-induced spirituality suffers from any number of serious drawbacks. First of all, spiritual states when forced artificially can result in nonspiritual states of mind, such as psychosis and delirium. In inexperienced and unprepared subjects, they are likely to cause severe anxiety and panic.

This is the least desirable path for the spiritual aspirant. Forced spirituality, in the absence of supportive knowledge, character structure, and lifestyle, will not be spirituality. In fact, it more often leads to fanaticism and cultishness.

Freedom is the capstone of spiritual experience. Use of drugs for spiritual enhancement can produce the opposite effect. Physical and psychological dependency on alcohol and cocaine are well known. The argument has been put forward that dissociation-producing drugs, such as marijuana and the hallucinogens, are not very addictive. While there is some truth to this statement, there are reports of severe addiction to both. My Native American friends and colleagues have spoken to me, for example, about addiction to peyote via participation in the peyote ritual.

Drugs are believed to bring about their behavioral effects by selectively influencing neurotransmitter mechanisms. These processes are subject to tolerance development. In other words, with repeated use, the same dose may cease to have the same effect. This, in turn, triggers the ever-increasing need to increase the dose to produce the same effect. After the system has become accustomed to having a high dose of the drug on board, abrupt discontinuation is likely to produce withdrawal symptoms. This is especially the case with alcohol and other disinhibiting drugs. Alcohol withdrawal can be unpleasant and may lead to serious medical complications, including death. Fear of withdrawal often reinforces the addictive process.

Violent outbursts are not uncommon, both after alcohol and cocaine intoxication. Violence toward self and others has also been reported after LSD. As we have seen, primordial consciousness is undifferentiated. As life evolved, it enlarged and diversified into constructive and destructive elements, both being essential components of the phenomenal world. There can be no creation without destruction, and the agency that mediates construction will also have to put in place the destructive element. However, neither construction nor destruction is relevant to undifferentiated consciousness. While the brain stem mediates neither construction nor destruction, at the midbrain level both are well developed. The hypothalamus, which controls nutrition and procreation, also regulates rage and violence. Release of these brain structures from inhibitory control can unleash violent impulses, instead of serenity and tranquillity.

Drugs are ingested for their desired effects on the brain. Unfortunately, we do not have techniques to administer the pharmacological agent selectively to targeted brain regions. Drugs given by mouth and by injection are delivered to various organs of the body, where most have effects in addition to the effects on the brain. This gives rise to side effects.

Use of these agents that were part and parcel of religious ceremony in olden days in India fell into disfavor in later days. It is important to note that soma (and perhaps marijuana) use was highly regulated and controlled, and it was used during elaborate ritualized practices and procedures. Nobody took soma privately at home or in the company of peers to make merry.

Although later scripture made reverential mention of soma, none recommended its continued use. In fact, all better-known religious traditions in India revile the use of drugs and inebriation to accomplish spiritual enhancement.

After his enlightenment, the Buddha gave his first sermon at Sarnath to the five spiritual aspirants he had associations with in his earlier days. He specifically spoke against the use of intoxicants. Freedom from inebriation is mentioned as one of the five precepts of Buddhism (Pañca-śīla).

Śāṁkara, much more vocal in his rejection of intoxicants, says: "A Brāhmaṇa shall ever avoid wine. They should pour boiling wine into the mouth of a Brāhmaṇa who drinks liquor" (Brahmasūtrabāṣya, III, 4–30.)

Chāndogya Upaniṣad equates drinking alcohol with the most heinous of other forms of misconduct: by those who steal gold, commit adultery with his teacher's wife, and murder a Brāhmaṇ (V, I, 9).

All spiritual teachers and religious leaders of repute in India have considered the use of drugs as a means to spirituality as self-defeating and counterproductive. I do not know of any respected teacher or leader in India who does not eschew drug-induced states, however close or similar they may be to the naturally occurring spiritual state of mind.

With the passage of time, a variety of techniques for enhancing spirituality that were less troublesome and tortuous and had more satisfying and enduring results were developed. These approaches effaced and expunged the use of drugs from present-day spiritual practices, and drug use exists only as a vestige of the past in isolated anomic groups.

Beauty and Beatitude

Yoga: Consciousness as Refuge

According to the Taittrīya Upaniṣad, the *kośas* that encircle the central core of consciousness represent successive stages in an evolutionary chain much like the growth rings seen in a tree trunk. Consciousness represents the primordial Absolute from which matter was derived. Life arose from matter. With additional growth, life enlarged and expanded to form the mind that subsequently transformed into intelligence. The Upaniṣad accords with the theory of evolution, in a broad sense.

Paul D. MacLean, the famous neurophysiologist, showed that the human brain has three independent though interconnected layers, representing milestones of brain evolution (Figure 7.1). The first and most primitive reptilian component deals with preservation of self and the species. Such behaviors as hunting, homing, mating, fighting, and the like are contained here. The next layer is the paleomammalian component that deals with such behaviors as parenting and play. The most highly developed component is the neomammalian part. More sophisticated functions involving components of intelligence—problem solving and memorization, language, reasoning, and so forth—are represented here. MacLean did not delve into the distant past before the reptiles. Since the reptilian brain was not created de novo from nothing, that which preceded it is likely to have left its mark, too.

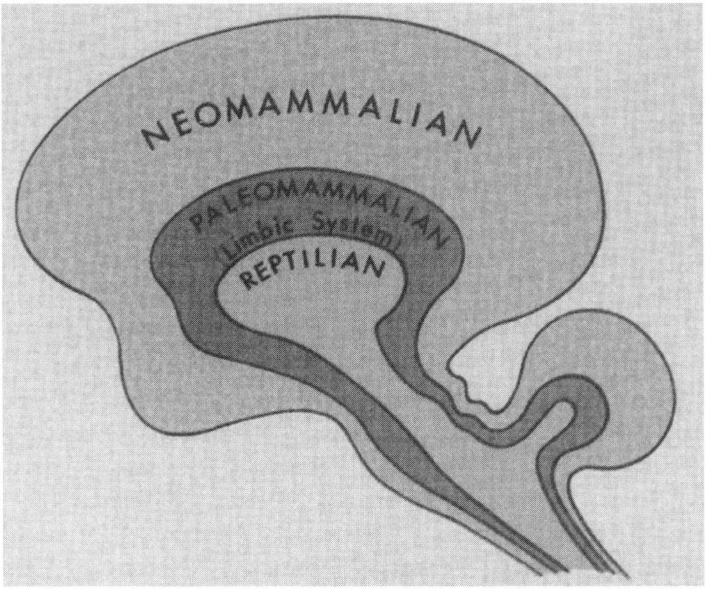

FIGURE 7.1 ***The triune brain.*** *According to Paul D. MacLean, the human brain has three separate but closely interconnected layers that reflect our ancestral relationship with reptiles, early mammals, and later mammals.*

The Upaniṣadic concept of evolution sharply and decisively disagrees with scientific theories of evolution, however, on one important point. The Upaniṣads do not see the unfolding of the cosmic saga from the big bang to the rise of intelligence as a mechanistic, self-perpetuating process. Several scientists attribute the inexplicable but key links in the long chain of events that undergird evolution to chance. The Upaniṣadic sages consider the cosmic cycle neither accidental nor coincidental. It is an intelligent process aimed at a determinate goal. Human intelligence, which is but an infinitesimal fraction of this mighty, impersonal wisdom, might not be able to grasp its scheme and scope in its entirety. However, since the brain is self-aware through introspection, it can retrace its evolutionary steps, subjectively, all the way to the starting point, the uncreated Absolute.

According to Indian thought, the primordial unmanifested Absolute and transcendental consciousness are one and the same. Mind, as we have seen, is simply a derivative of consciousness—the Absolute—with characteristics of the mind being, therefore, de facto characteristics of the Absolute (Muṇḍaka Upaniṣad, 1, 1, 8). According to the second-century Buddhist philosopher Nāgārjuna, "Nothing of *saṁsāra* [phenomenal existence] is different from *nirvāṇa* [transcendental reality]; nothing of *nirvāṇa* is different from the phenomenal world" (Madhyāmaka Kārikā, XXV, 19). This can be confusing, since it was stated earlier that the Absolute is predicateless. This can be resolved by understanding that although it is devoid of quality, it has the potential for quality. All the qualities of a human being are carried in the human gene, although the gene itself does not resemble a human. The transcendent consciousness generates the rest of the mind, bound in time and space. The newly formed layers of the mind cloak the consciousness within. As we have seen, this phenomenon of concealment and distortion is called *māyā,* and its acceptance as real, *avidya.* Enlightenment entails transcending time- and space-bound layers of the mind to experience the uncontaminated consciousness.

Derived from the root word *yuj*—to unite—the term "yoga" denotes the confluence of the transcendental consciousness and the mind. Although consciousness is the basis for the mind, most individuals identify with the mind where consciousness is distorted and contaminated. In other words, concept of self is based primarily on the mind and not consciousness. Thus, ordinarily, the mind eclipses consciousness. In enlightenment, the emphasis shifts from the mind to consciousness, from the phenomenal to the noumenal, and from the carnal to the spiritual. The enlightened see themselves primarily as consciousness and recognize the mind as an evanescent, nonessential appurtenance. Yogic techniques in general facilitate this transformation.

Although the entire mind is derived from consciousness, the conscious element is not evenly visible everywhere. *Māyā* and *avidya* are heavier in some areas than the others. The yogas represent some of the better-known windows of access, across the veil of *māyā*.

Since antiquity, enlightenment seekers resorted to many paths and practices and, in India, they all are subsumed under the caption of yoga. Songs and sacraments, fasting and feasting, praying and praising have all been found useful and beneficial. The absence of a single, well-established, correct way is widely acknowledged. "As birds in flight and fish in the ocean leave no traces of their paths behind, people who found enlightenment also do not leave any tracks behind" (Mahābhārata, Śāntiparva, 181, 12.) The Buddha's last words were, "Work out *your own* salvation" (Mahāparinibbaṇa sutta, VI, 10).

The main message from the Upaniṣads and the Buddha is that each person has to find his or her own way. However, Hindu scripture does provide some broad guidelines. The most celebrated account of yoga can be found in Bhagavad Gītā, where Lord Kṛṣṇa expounds three paths to achieve yoga to his devotee, Arjuna. Bhagavad Gītā, referred to as Gītā in short, is part of Mahābhārata, the larger of the two epics.

Gītā, the essence of Hinduism, is the distillate of the Vedas and Upaniṣads. It reveals the divine logos and the love of the Creator for the creations.

Mohandas Gandhi wrote in *Young India* (August 6, 1925), "When doubts haunt me, when disappointments stare me in the face, and I see not one ray of hope on the horizon, I turn to the Bhagavad Gītā and find the words to comfort me; and I immediately begin to smile in the midst of overwhelming sorrow."

Mahābhārata is a rich storehouse of stories within stories, interesting, entertaining, and illuminating to Indians of all ages, castes, creeds, and religions. It is believed to be the Āryanized version of the bardic songs about the great war waged between two branches of one royal family, the Bharatas, leading to the great battle of Kurukeṣetra.

Since they were blood relatives, the armies that lined up on either side were made up of kith and kin. Of the two warring parties, five brothers, including the archer par excellence named Arjuna, represented the good, and the other, made up of 100 brothers, the evil.

Both groups approached Kṛṣṇa, a powerful king and their revered relative, for military support. For the devout, Kṛṣṇa was the divine in human form, and he is one of the most revered deities in India. Kṛṣṇa, eager to appease both, offered to join one side in person and to give his army to the other. The villainous group, greedy and unscrupulous, wholeheartedly accepted the services of his soldiers while the other group, humble, devoted, and just, was more than pleased to have Kṛṣṇa join their side. Kṛṣṇa added one more condition: He would just be an observer and adviser, but not an active participant in the actual fighting.

On the day of the great battle, before the first blow was stuck, the archer Arjuna asked his charioteer Kṛṣṇa to drive between the two armies. He was flabbergasted by the sight of the many friends and relatives on the other side and horrified at the thought of harming, not to mention annihilating, them. Although Arjuna knew that the battle was inevitable, his participation essential, and the cause noble, he could not bring himself to injure and kill the near and the dear. Confused, frustrated, and anxiety-ridden, Arjuna discarded his mighty bow and declared his unwillingness to take another step forward. In fact, Arjuna provided the first detailed description of an anxiety attack ever recorded: "My limbs have grown feeble. My mouth is parched. My whole body shakes and shivers. My hair bristles. My skin is on fire. My bow is slipping from my cold, clammy hand. I am dizzy and my mind is befogged" (I, 29–30).

The eighteen chapters of Bhagavad Gītā contain Kṛṣṇa's consolation and counsel. Arjuna wanted to know how one can be freed from the dichotomous phenomenal world of love and hate, creation and destruction, good and evil. Phantasmal *(maya)* though it may be, the mental anguish, pain, and suffering are as real as can be. How can one escape the tumult and find inner tranquillity and peace?

The answer to Arjuna's question is to discard the phenomenal world and to take refuge in the sanctum sanctorum of undivided consciousness. The problem is how.

Kṛṣṇa, in response, brought up the concept of yoga. Salvation may be accomplished through many means, depending upon the person's station in life, constitution, temperament, inclination, and personality. To reach the ocean, different rivers take different courses; none is better, preferable, or superior.

Several authorities hold that Kṛṣṇa spoke of three yogas: the Jñāna yoga—union through knowledge, Bhakti yoga—union through devotion, and Karma yoga—union through conduct. There are, however, ample references to a fourth, Rāja yoga—the royal path. The four should not be seen as mutually exclusive; in fact, they overlap to a great degree.

Union Through Knowledge

Jñāna is knowledge. Spirituality and intellect are not antithetical, as some would have it. Numerical reckoning, upon which science is based, is just one type of intellection. In fact, the best-known thinkers in India and elsewhere were highly intellectual, though not scientific. Jesus and the Buddha were intelligent but in a different way from Einstein and Darwin.

Early thinkers, unaided by the tools of science, arrived at the same conclusions as modern scientists concerning the unreality of the phenomenal world. The scientist reduced matter to atom and further to subatomic particles to conclude that ultimately all matter is nothing but different configurations of energy. The sage, however, through empirical reasoning aided by intuitive insight, declared that the material world is simply a manifestation of a transcendental Absolute. Many scientific discoveries of import were serendipitous, and noted scientists have acknowledged the importance of intuition in their work.

Knowledge in the usual sense means understanding of the manifested world. Knowledge of that which manifests the world will have to be the locus classicus. Science, notably mathematics, has discovered elegant equations that describe the cosmos. These equations betoken the intellect that launched them and evince the genius of creation. As the Indian peacock-egg hypothesis states, the results are manifestations of what is latent in the cause. The egg contains the bird's beauty, in a dormant form.

The factual intellect of the scientist and the intuitive wisdom of the sage are both created elements and derivatives of the same primordial, incorporeal wisdom. The wellspring of all wisdom is the Absolute that exists within the human mind. According to the Bṛhadāranyaka Upaniṣad, "He who knows self knows everything" (1, 7, 1).

The Hemispheres of the Brain

Both the metric pattern of science and the nonlinear, holistic approach of the thinker are dependent upon brain mechanisms. These sophisticated mechanisms came into being in humans much later in evolution as the Absolute transmuted into formed elements.

As the brain continued to expand and enlarge, it took on newer functions, more complex and complicated. Different brain regions mediated the newly emerged diverse functions. The burgeoning of functional capabilities necessitated rapid, morphological expansion of the brain, which was limited by the bony cranium. The skull afforded the brain protection from injury, but it also restricted its expansion. The cortical surface of the brain folded upon itself to increase its surface area. As a result, the human brain has a rugose, corrugated surface.

The human brain has two roughly symmetrical halves with a connecting bridge. This bilateralism in brain morphology can be traced all the way back to flatworms. In animals with symmetrical body parts and brains, the contralateral half of the brain is wired to the other half of the body. As the brain acquired elaborate and intricate functions, some were consigned to one hemisphere while others went to its counterpart. Interhemispheric functional asymmetries have been found even in birds and bats. In canaries, it is the left side of the brain that sings. In dolphins, the right hemisphere responds to simple commands, while the left hemisphere deals with more complex ones. The functional asymmetry is predictably more marked in the human brain with its myriad functions.

Some 2 million years ago, in *Homo habilis*, the two hemispheres may have undergone marked functional reorganization with the development of brain regions that mediated speech. These regions continued to grow and develop further in *Homo erectus* (1.3 million years ago) and reached the present level in *Homo sapiens neanderthalis*, about 100,000 years ago. Spoken speech became well established in *Homo sapiens sapiens* approximately 40,000 years ago. During the last 10,000 years, reading, writing, and math came into being. The left hemisphere that carried out these functions became dominant. As evolution unfolded further, the two hemispheres acquired functions of greater complexity, intricacy, and sophistication.

Functional differences between the two hemispheres received firm experimental verification when a neurosurgeon, W. P. Van Wagenen, severed the corpus callosum, which joins the two halves of the brain, to prevent the spread of epileptic seizures from one hemisphere to the other. The procedure was not accompanied by any obvious deterioration in the patient's mental condition and did relieve the epilepsy to some extent.

The surgical procedure thus developed allowed an experimental psychologist, Roger W. Sperry, and his associates to conduct an elegant series of research projects on the functional differences between the disconnected hemispheres. For his epoch-making findings, Dr. Sperry received the Nobel Prize in physiology and medicine in 1981. The following years saw a burgeoning mass of experimental data on the functions of the dominant and nondominant hemispheres. Some findings were replicated repeatedly and became firmly established, while others remained speculative and weaker.

Comprehension and expression of language, both spoken and written, are firmly established as left hemispherical functions in approximately 99 percent of right-handed individuals. Of the sinistral and ambidextrous, around 60 percent still have left hemispheric dominance for language. In the remainder, language is bilaterally represented. The left hemisphere is active in programming of complex sequences of movement. It is also involved in the preservation of body image. Logic, reasoning, temporal and sequential information processing, attention to details, and so on are dominant hemispheric functions, in addition to spoken language. Music, art,

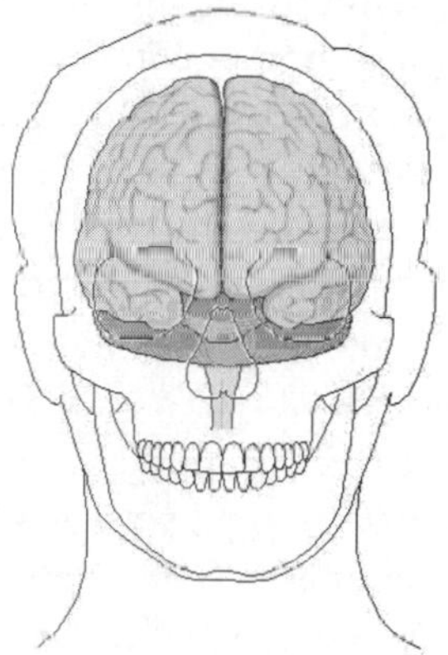

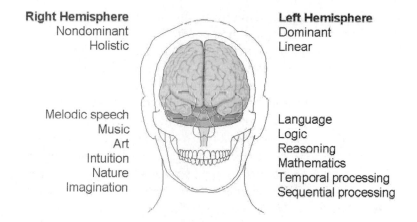

FIGURE 7.2 ***Functional organization of the two cerebral hemispheres.*** *In right-handed people, the left hemisphere is dominant.*

three-dimensional forms, imagination, emotional and melodic speech, intuition, perception of environmental sounds and scenes, and social and emotional nuances are mostly nondominant functions (Figure 7.2).

The right hemisphere has the capacity to relate to nature—the song of a bird and the clamor of a thunderstorm. Damage to the right side of the brain interferes with the ability to recognize music. The left hemisphere, on the other hand, is concerned with a temporal-sequential organization and recognition-analysis of details. The dominant hemisphere cannot and does not see the whole. It simply sees the components. The right brain, on the other hand, can bridge the gaps, add details, and perceive the whole. For example, the right brain is more proficient at recognizing faces than the left.

Numerical logic and inductive reasoning are, therefore, left-hemispheric functions. It is the home of all sciences. Temporal and spatial segmentations and their tracking and processing are left-hemispheric functions. The right, less attentive to individual components than to the collective whole, is more involved in arts. The scientist who sorts out resides in the dominant hemisphere, while the artist who sums up has the nondominant right hemisphere as home. Hard facts, particularity, and inductive logic drive the dominant hemisphere, while intuition, abstraction, and imagination dictate the nondominant hemisphere. Daydreaming and reverie arise from the right hemisphere, while dealing with cold facts and making calculated decisions come from the other hemisphere.

Knowledge, including that of self, requires participation of both logic and intuition. The Buddha in his quest for enlightenment came to the conclusion that the right path was the middle path that avoids extremes. His teachings do provide some logical explanations, although he abhorred excessive intellectualization. The main appeal of the Upaniṣads is the use of logic and reason, based on empiricism, and not freestanding, blind faith. Free spirit of inquiry unfettered by dogma is an outstanding feature of both Buddhism and the Upaniṣads.

The lower knowledge that deals with the space- and time-bound, manifested world is *vijñāna,* while *jñāna*, true knowledge, deals with the uncreated element. While the former deals with sparks, the latter acquaints us with fire. *Māyā* that cloaks and conceals the undivided reality creates the phenomenal world. As was pointed out earlier, the concept of *Māyā* is derived from the root word *mā*, meaning to measure, a dominant hemispheric function.

Factual knowledge, in general, informs us about what is unreal, but not what real is. One has to recognize the unreality of the phenomenal world before delving further.

Scientists with their arcane equations have shown us that space and time are relative and variable and that, therefore, phenomena dependent upon it are ephemeral. But science does not tell us about the nonephemeral and perdurable.

Science has traced the origin of the universe to the big bang but cannot tell us what caused the big bang and what preceded it. Since time was nonexistent before the big bang, the question is probably redundant, in a material sense.

Factual knowledge is mediate, dependent upon perception, but the intuitive insights are direct and immediate. Intuitive knowledge, which every living person is aware of at some level, involves merger and union: It is quite distinct from learning from a book.

Union Through Devotion

Chāndogya Upaniṣad illustrates this vividly in chapter 7, section 1. The learned sage Nārada conversant in all Vedas and *śāstras* (sciences) goes to Sanatkumāra, the five-year-old innocent child, for true knowledge. The point here is that true knowledge does not require intellectual sophistication. Together they consider different aspects of the manifested world and related phenomena and dismiss them one by one. Finally, they conclude that true knowledge is knowledge of self.

How can the self be known? Muṇḍaka Upaniṣad says, "Brahma veda bhamaiva bhavati" (III, 2, 9); to know the Absolute is to become the Absolute. Śāṁkara confirmed this view in Vivekacūḍāmaṇi: "As milk poured into milk, oil into oil, and water into water mix and become uniform, so the knower of the Absolute becomes one with it" (verse 566).

Bhakti yoga (devotion) is based on belief in a primordial Absolute, vibrant with life. The Absolute from which life evolved cannot be passive, detached, and lifeless. Enlightenment, according to the Upaniṣads (but not the Buddhists) is therefore not one-sided; it is a bilateral process involving both the creator and the created. According to the Muṇḍaka Upaniṣad, "This self cannot be attained by instruction nor by intellectual power nor even through much hearing. He is to be attained by the one whom the Self chooses. To such a one the Self reveals his own nature" (III, 2, 3).

Love is the yearning for union. Since all creations are manifestations of a primordial Absolute, all love is love for the Absolute. In Madyāmaka Kārika, Nāgārjuna, the Buddhist philosopher, points out that in reality, all activities are expressions of one's longing for the Absolute *(dharmaiṣaṇā)* that is one's ultimate nature. The attraction we feel toward both animate and inanimate objects around us is a reflection of our longing for yoga. Bṛhadāranyaka Upaniṣad says that the love one feels for one's spouse, children, wealth, cattle, people, the world, and even gods, in reality, is an expression of our desire to be a single, whole one (IV, 5, 6). The serenity we experience while gazing at the starlit night sky, the bliss we feel when looking at a rose in full bloom, and the joy that fills us while caressing a pet are manifestations of the yoga that took place.

The rosy warmth of romantic love, the Upaniṣad says, is also the bliss of union between two individuals previously separated by *māyā*. Love transcends time and space, and it ushers in the bliss of spiritual union. According to Bṛhadāranyaka Upaniṣad, "As a man in the embrace of his beloved wife knows nothing within or without, so the person when in the embrace of the intelligent self [the Absolute] knows nothing without or within" (IV, 3, 21).

Paul expressed comparable views on marriage in his Letter to the Ephesians: "Therefore as the church is subject unto Christ so let the wives be to their husbands in everything. Husbands, love your wives even as Christ also loved the church and gave Himself for it" (Ephesians 5.24, 25). Christ's relationship with the church was unquestionably spiritual and not carnal. The latter is yearning for gratification of a narrowly defined self and not for union. Love is yearning for yoga, and the associated bliss is the result of its accomplishment.

Christianity is often hailed as the religion of love. In response to an inquiry about the greatest commandment, Jesus replied: "Thou shalt love the Lord, thy God, with all thy heart, with all thy soul, and with all thy mind. This is the first and great commandment. And the second is like unto it, Thou shalt love thy neighbor as thyself. On these two commandments hang all of the law and the prophets" (Matthew 22.36–40). This is yoga, from the Indian perspective.

A passage from the Gospel according to John clearly illuminates and illustrates the concept of Bhakti yoga, or union through devotion: Jesus prayed, "Neither pray for these alone, for them also which shall believe in me through their word; that they also may be one; as thou, Father, art in me, and I in thee, that they also may be one in us; that the world may believe that thou hast sent me" (17.20, 21).

Jesus conceptualized humans as children of God. Love of God for his children is central to Christianity. This overwhelming affection between the created and the Creator is the touchstone of Bhakti yoga. In Bhagavad Gītā, Lord Kṛṣṇa tells Arjuna, "Even a leaf, flower, fruit, or water is acceptable to me when offered with love and a pure heart" (Chapter 9, verse 26).

Devotion is unlikely to be a function mediated by the linear and particular dominant hemisphere. Papers on functional hemispheric differentiation usually list religion under right hemispheric functions. Yoga means union and, therefore, is more likely to fall under the holistic right hemisphere than the fissiparous dominant hemisphere.

Prayer is the universal device for holding communion with God. There would appear to be two types: solitary prayer and group prayer. There are a number of similarities between solitary prayer and meditation. In prayer, the devotee longs for communion with God; in meditation, the sage seeks union with a nameless and formless Absolute.

The notion that God is within the human mind is firmly established in Hinduism. This concept is by no means unique to Hinduism. According to Luke, when the Pharisees asked Jesus when the Kingdom of God would come, he said: "The Kingdom of God cometh not with observation: neither shall they say, Lo here! or Lo there! for, behold the Kingdom of God is within you" (17.20–21). According to John, "Even the Spirit of truth; whom the world cannot receive, because it seeth Him not, neither knoweth Him: but ye know Him; for He dwelleth with you, and shall be in you" (14.17).

In pantheistic Hinduism, God is anthropomorphic. However, most Hindus, especially the more philosophically minded, would acknowledge that ultimately, they are one and the same. According to Darśanopaniṣad, "The vulgar look for their gods in water, men of vital knowledge in celestial bodies, the ignorant in wood or stone, but the wise see the supreme in their own self."

Quiet, peaceful, and picturesque surroundings are conducive to prayer. Jesus went to the mountains before he started preaching and to the Garden of Gethsemane immediately prior to his arrest. He often withdrew from his disciples and prayed in isolation. Freeing the mind from the rough-and-tumble of worldly concerns is considered essential. A pure mind and a clean heart facilitate prayer.

Group prayer is more closely tied to religious rites and rituals. A number of activities, including singing and chanting, often accompany this type of prayer. There is a close association between art and prayer, especially group prayer. The famous Indian philosopher Radhakrishnan quotes the Upaniṣadic passage, "Art is that which brings one into contact with the divine."[1]

Many religious texts are written as lyrics or poems. Psalms of the Old Testament, the Vedas, and the Holy Qur'an are striking examples of this. The Qur'an, I am told by my Arab friends, is the crown jewel of Arabic literature. It is remarkable not only for its elegance and poetic beauty but also for its syntactic perfection. Rendering of the Qur'an is a highly evolved art requiring considerable talent and training. The Vedas were anthologies of hymns developed by bard-priests to render during rites and rituals. In the Ṛg Veda, the term "sage" *(ṛṣi)* is interchangeably used with "poet" *(kavi)*. Sāma Veda reformulated Ṛg Vedic verses and added pauses, prolongations, and inflections to give it a distinct musical quality.

Singing is an important part of Hindu religious rituals. Almost all Christian churches in the West have choirs, and religious music is an integral part of Christian ceremonies. Dancing is part of the ritual for Sufis, an esoteric faction of Islam. Drumming and dancing are practiced by Hindus, North American native cultures, and different African and Australian tribes. Some of the world's best pieces of art, both paintings and sculptures, adorn churches, temples, and mosques. Aesthetic experiences clearly enhance religious ecstasy.

Lord Śīva, of probable Indus Valley origin, is believed to be the creator of the three major art forms—music, dance, and drama (Figure 7.3). He created rhythm with *ḍamaru,* the percussion instrument that adorns his hand. He composed the first five *rāgas* while his consort, Pārvati, produced the sixth.

Rāgas are fundamental melodic types that serve as forms for the creation of different types of compositions. *Rāgas,* with an ascending and descending scalar structure, embrace a variety of emotions. Certain *rāgas* are to be sung only in the morning, while others are rendered only during evening hours.

FIGURE 7.3 *Śīva, one of the three principal Hindu gods, is believed to be the creator of music, dance, and drama.*

Rhythm, in addition to its obvious relevance to music, is also related to prayer. Most religions have short phrases; for example, Christians utter "amen" and "hallelujah" during prayer. This tendency is much more pronounced in Hinduism. The sounds that are reverentially chanted are called mantras. Mantra has its etymological roots in the word *manāna,* meaning "to think." Mantras are that by which the contemplation of God is attempted.

Music has a unifying effect. Singing, especially when part of a religious ceremony, brings (and binds) the devotees together. The musician and the audience equally share the experience of music. Both the producer and the receiver immerse in the common experience of aesthetic pleasure. Most artists love to share their art, and most aesthetes like to listen to music in the company of people they love. Indians use the Sanskrit term *tanmayībhāva* to describe total involvement, immersion, in art.

There is little difference between the artist and the aesthete; both are essential. Both the flame and the cinder are interdependent in the creation of fire. Absorption in aesthetic appreciation softens and melts temporal and spatial barriers and causes the loss of the ego in the current of transcendence. It is conducive to "yoga," or merger.

Beauty and the Arts

A number of investigators have studied the brain mechanisms associated with aesthetic emotion and appreciation. Although we have some ideas about the mechanisms involved, much remains unknown.

Music clearly has two components: the technical and the emotional. There is a great deal of technical information useful (but not indispensable) to a good musician and perhaps even to the aesthete. As we have seen, while one half of the brain, the dominant half, mediates factual knowledge, the other one deals with the subtle emotional aspects.

The Wada test is employed to identify the hemisphere that mediates various functions, most commonly language. The effect of selective inhibition of each hemisphere, on the function of interest by the injection of an anesthetic into the corresponding carotid artery (Figure 7.4), is studied in this way. Singing clearly involves use of language and sequential use of words. The dominant hemisphere mediates language, and as might be expected, when the dominant hemisphere is anesthetized, difficulty in singing words is produced. When the anesthetic is injected into the nondominant hemisphere, severe melodic distortion is caused.

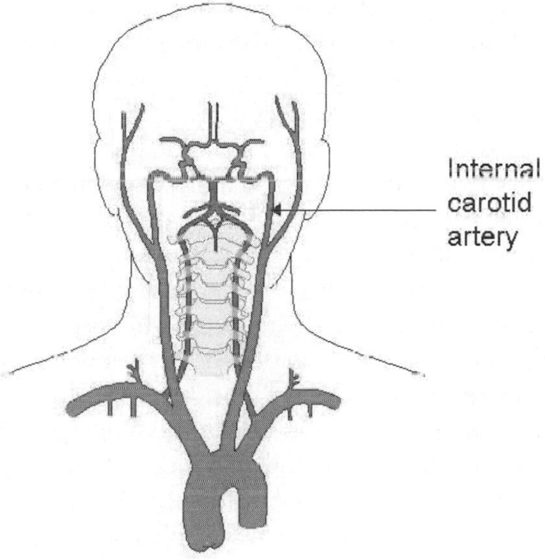

FIGURE 7.4 *The internal carotid artery of one side perfuses the corresponding cerebral hemisphere. In the Wada technique, one injects a barbiturate into one carotid artery to anesthetize the cerebral hemisphere on that side.*

Appreciation of music involves audition, and auditory processing is a temporal lobe function. Damage to the temporal lobe causes auditory processing defects. While the linguistic component of songs is a dominant hemispheric function, the melodic aspect is dependent upon the nondominant hemisphere.

Surgical resection of the nondominant right hemisphere, usually for the control of epilepsy, results in impaired tonal processing, melodic discrimination, perception of the pitch, and memorization of the pitch. The linguistic left hemisphere has a role in the appreciation of hearing music with lyrics. However, enjoyment of instrumental music is predominantly if not exclusively a function of the nondominant hemisphere.

In one research project, epilepsy patients with their temporal lobes surgically removed were shown the text of a familiar song and they simultaneously heard it sung. The results indicated that patients with right temporal-lobe resection performed significantly worse on both tasks than either patients with left-temporal lobe lesions or normal subjects. The results suggest that the right temporal lobe specializes in perception and imagery of songs. The temporal lobes have close and intimate associations with the frontal lobes and the right prefrontal lobe was found to be associated with pitch discrimination.

Degenerative diseases of the dominant hemisphere affect linear functions, while those involving the nondominant hemisphere impair holistic functions, including music. In one report, a singer in a chorus experienced loss of musical abilities and aptitudes with the progression of right frontal and temporal lobe damage. She had completely lost the capacity to sing or hum a tune, even the most familiar ones, but did not have any difficulty with her speech or perception. She was also unable to identify common melodies by name in spite of an intact memory and ability to name common objects. In addition to problems with the sense of rhythm, she could not discern or reproduce rhythmic patterns.

Dr. William H. Wilson and I conducted an experiment in which subjects were paid to perform arithmetic tasks and, on another visit, for which they were not paid, to listen to music. The arithmetic tasks increased brain activity, as expected, in the dominant hemisphere, whereas music increased it in the nondominant hemisphere. Other investigators have shown that analytical processing of music by musical experts engages the dominant hemisphere, while uncritical appreciation by the music lover involves the nondominant hemisphere.

While recognition of complex rhythms may in part be mediated by the dominant hemisphere, studies involving the Wada test suggest that injection of the anesthetic into either carotid artery does not obliterate the basic sense of musical rhythm. This would make sense, since even animals with much less cortical development seem to be sensitive to rhythm, as evidenced by the wagging tail of a dog and the cobra swinging in consonance with the snake charmer's movements. In Indian festivals, it is a common sight to see elephants fanning their ears to the beat of drums.

Pulsatile activity that forms the basis for rhythm is fundamental to life. The electroencephalogram, with the best temporal resolution of all techniques available for monitoring brain activity, shows pulsatile activity of the brain even at the most basic level. Cessation of this rhythmic pattern means death.

Oscillations go even further back beyond life. Light is composed of oscillations of energy particles. There is nothing in nature that is static. Everything vibrates, pulsates, and oscillates at different frequencies.

Between the ninth and eleventh centuries A.D., Kashmir Śaivites, an esoteric group of Indian theologians, claimed that pulsatile activity was the first created element. Transcendental consciousness, anterior to the created element, is devoid of movements, including pulsatile ones. Movement requires time and space, both of which are created elements. Vibrations that characterize empirical consciousness are referred to as *spanda*, which literally means "pulsation." As creation unfolded, cycles of widely varying time integrals appeared—from the cycle of cosmic creation and annihilation, lasting billion of years, to the pulsatile movements of subatomic particles of microsecond duration. The phenomenal world is composed of a plethora of pulses of varying lengths and durations.

Ṛta, or rhythm, according to Ṛg Veda, is a fundamental property of the manifested world. Rhythm is also the basis for music and all dance forms. Although in the West dance has little association with religion, in India the two are very closely related. Images of Śīva commonly show him in the performance of the *thāṇḍavam*, a dance form symbolizing the fury of cataclysmic destruction. The Ṛg Vedic goddess Saraswatī is the patron of all art, especially music (Figure 7.5).

Rhythm has a captivating effect on humans and animals. The hypnotizing effect of rhythm is the basis for many tribal dances. Rhyme is the life and soul of poetry.

Having a subject focus on the swinging movements of a watch on a chain induces hypnosis, a dissociative state, although little is known

FIGURE 7.5 Saraswathī, a Hindu goddess, is the patron of art. Images of her usually show her with veena, a stringed musical instrument.

about the associated brain mechanisms. Photic driving is a well-known phenomenon where the frequency of the subject's EEG takes on the frequency of a pulsatile strobe light that he or she watches.

It is conceivable that a similar mechanism might be in operation with the beats of a drum as well. The rhythm of brain activity may align itself with the beats. Total absorption with rhythmic beats of the drum will compromise the brain's ability to attend and respond to other stimuli, internal and external. This would result in cancellation of most cognitive activities and total absorption into the rhythm of the drum.

Most people can identify with the experience of loss of time and space sense while buoyed up in the blissful experience of music. Anecdotal reports are available on the slowing of EEG during total immersion in music. To the best of my knowledge, there are no research reports on the subject.

We do know that music is soporific. Dreamless sleep is accompanied by EEG slowing. In addition to music, rhythmic tapping and swinging movements of the crib help the infant fall asleep. The mood states associated with dreamless sleep and music, though different, seem to represent two ends of a continuum.

Music requires cerebration, but sleep does not. Music ushers in sleep; one blends with the other. Music silences the logical dominant hemisphere and activates the holistic nondominant one. The rhythm silences both cerebral hemispheres and plunges us into greater depths to the subcortical parts of the brain, probably the brain stem. This is indicated by slow, rhythmic EEG waveforms. The less contaminated the consciousness, the more tranquil and blissful the associated state of mind will be.

Like music, many celebrated works of painting and sculpture are on religious themes. Michelangelo's paintings and the statue of the madonna at St. Peter's Cathedral in Rome are well known. In India, art is even more closely associated with spirituality.

Maṇḍalas and yantras are diagrams with symbolic significance that enable the aspirant to transcend the material world. These diagrams possess a hypnotizing effect similar to that of slow drumbeats. Spiral images that perform a similar function are also used to induce hypnotic trances. The characteristic architectural styles of Hindu temples are fashioned after mandalas.

Sri-chakra, shown in Figure 7.6, is one of the best-known yantras. The diagram is visualized as an elaborate mansion with the central point "*bindhu*" representing transcendental consciousness (usually symbolized as the mother-goddess). It is the point of cre-

FIGURE 7.6 ***Sri Chakra.*** *It is seen as a symbolic map of both the universe and the human mind. Outer layers represent created elements, and the central point, pure consciousness.*

ation where the formless conscious element transmutes into matter cast in space and time. It is cloaked in nine veils, representing the obstacles the spiritual aspirants must overcome for Enlightenment, or union with the goddess. The outer walls, made of three lines, provide four entrances, representing the four yogas. Layers of sixteen and eight lotus petals constitute the second and third veils. The remaining six are made of the outer boundaries of the interlacing triangles within and the point in the center. Every line and shape has a name with special symbolic meaning.

Sri-chakra may also be seen as a map of the human mind. The outer layers represent perception of the material world and the inner ones such related mental functions as cognition and memory. In the innermost sanctum, matter fades into spirit. The central core is pure, unsullied consciousness.

The Sri-chakra has been traced back into the mystic haze of the very distant past. Many believe it came from the Atharva Veda, with close ties to the Indus Valley civilization. In spite of its current use in orthodox Hinduism, it has definitely had closer associations with Tantrism, a heterodox movement. It is worshiped in several important Hindu temples, and mystics and occultists revere it. Many who admire it primarily for its grace, elegance, and beauty keep it in their homes and places of work for good luck.

Sri-chakra, when taken apart, is nothing but a matrix of interlacing lines. It acquires beatific meaning and spiritual significance only when seen as a whole.

Appreciating Sri-chakra is clearly a right hemispheric function. Patients with right hemispheric injuries lack spatial coherence in their drawings. They draw appropriate parts, but the parts show inappropriate placement in relation to one another, and the three-dimensional quality is weak or absent. The right hemisphere is also superior in face recognition. Inability to recognize familiar faces is a common symptom of right hemispheric injury. This hemisphere is better equipped to identify faces from blurred pictures. The right hemisphere is adept at extracting inner meaning and significance from simple diagrams. Thus, like music, the creation and appreciation of pictures, paintings, and sculptures are also carried out by the nondominant hemisphere.

Indian literature describes the Supreme as the first artist. It is the Creator who smiles in the flowers, sings through the birds, and caresses us with the nimble fingers of the breeze. "Rūpam, rūpam pratirūpam bhavati," says Ṛg Veda (6, 47, 18)—"All creations are but reflections of one primal image."

Art bespeaks the artist. The artist can often be recognized from the work of art—it bears his or her signature. A Beethoven or Picasso can be identified from his works. Similarly, creations speak of and for the Creator.

According to the peacock-egg hypothesis, the result is inherent in the cause. Thus, the beauty and splendor of the universe bespeak the nature of the creator. The ice that crystallizes into symmetrical patterns on the windowpane, the raindrops that bejewel the lotus leaf, and the neat formations of birds flying against the red blaze of the setting sun all tell us about the maker, the great artist.

Beauty is a fundamental characteristic of the cosmos, both as microcosm and macrocosm. The mighty galaxies and tiny snowflakes, the gigantic mammoth and the small microbe, and the sweeping oak tree and the humble grass underneath—all bear witness to the artistic finesse of the maker. The gentle whisper of the wind, the pitter-patter of rainfall, the gurgling of the brook are all pleasant to the ears.

All around us the mark of the Creator is obvious and evident. Beauty is that which produces bliss. Our instincts to reach out and touch beautiful articles and to possess them indicate our yearning for yoga, to unite with the Creator. In the depths of our minds, we feel the pain of separation and long for yoga, or reunion. Art is a vehicle.

Since Einstein, physicists and mathematicians in pursuit of a cosmic code, the single equation that could bring together all the fundamental forces of nature, recognize and acknowledge the importance of beauty. Einstein said, "Nature, though difficult to understand, ought at the root to be simple and beautiful."[2]

It might seem strange that mathematicians consider beauty as a dependable marker of mathematical concepts to be of cardinal significance. Paul Dirac, Lucasian Professor of Mathematics at Cambridge and 1933 Nobel Prize laureate in physics, felt that, "it is more important to have beauty in one's equations than to have them fit experiments."[3] Leon Max Lederman, the American physicist who won the 1988 Nobel Prize in physics explains: "As things stand now, it is too complicated. There is this deep feeling that the picture is not beautiful. The drive for beauty, simplicity, and symmetry has been an unfailing guidepost as to how to go in physics."[4] John Archibald Wheeler, the first American physicist involved in the theoretical development of the atom bomb and Blumberg Professor of Physics at the University of Texas at Austin, speculated about the ultimate mathematical equation that would explain the universe: "When we discover the equation or idea it will be so compelling, so inevitable, so beautiful that we will say how could it be otherwise."[5]

Art is, by and large, imitation of the first artist—the Creator. The seven Indian musical notes—Sadja, Rsabha, Gandhara, Madhyama, Pancama, Dhaivata, and Nisada abbreviated as Sa, Ri, Ga, Ma, Pa, Dha, and Ni—have been linked to the vocalizations of certain birds and animals. It should be mentioned that these notes are believed to be the forerunners of the seven basic notes of Western music. Sa stands for peacock; Ri represents bullock; Ga, goat; Ma, jackal; Pa, cuckoo; Dha, horse; and Ni, elephant. The claim that the seven sounds represent the seven colors of the rainbow has also been put forward.

There is also a close association between the *rāgas* and nature. The six principal *rāgas* represent the six seasons: Natnārayan, Summer; Megh, Rain; Pancham, Autumn; Sri, Dew; Bhairav, Winter; and Hindol, Spring. Some *rāgas* calm storms, tame wild animals, cause oil lamps to light up, and cure diseases.

In India, dances originated as temple art, and many are symbolic enactments of religious themes, usually drawn from the epics. Dances are rich in gestures with special meanings called *mudras*. Many of these *mudrās* and dance movements are imitations of birds, animals, and trees.

Rabindranath Tagore, the Indian poet and Nobel laureate writes in *The Religion of Man*, "Men must find and feel and represent in all their creative works Man, the Eternal, the Creator."[6]

Creation of art is often preceded by an intensely blissful, esthetic emotion Indians call *rasa*. What starts out as an ill-defined, imprecise, and vague sensation, after a variable period of gestation and formation, transforms into a well-developed and defined art form. The mood state from the conception to the delivery is uniquely blissful. In art, there is an intricate, bilateral interaction between the formless and the form, between the idea and the product, and between the abstract and the manifest. According to Rabindranath Tagore:

> *Incense craves to dissolve itself into fragrance and fragrance wants to remain enveloped in the incense. Melody finds itself articulated in the rhythm and the rhythm wants to go back into melody. Idea craves to be embodied in form and form seems to release itself in*

> *idea. The limitless seeks its intimate association with the limited and the limited craves to loose itself in the limitless. I know not whose logic it is in creation and destruction, that there is an increasing intercourse after freedom and freedom is always looking for a nest in bondage.*[7]

The interchange between the impulse to create and the creation, the abstract and the concrete, and the infinite and the finite parallels the interplay between the uncreated Absolute and the phenomenal world.

The words *satyam, śīvam,* and *sūndaram* (truth, grace, and beauty) are often invoked as characteristics of the Creator. These features characterize not only creation but destruction as well. The lightning bolt, the hooded cobra, and even the AIDS virus are exquisitely beautiful. Preoccupied with self-protection and preservation, we may not see this splendor as readily.

There can be no creation without destruction; they are as interdependent as the two sides of one coin. The Creator will therefore have to be the destroyer as well.

Good is meaningless without the backdrop of evil, and thus, God and the devil are mutually interdependent. The anthropomorphic polytheism of Hinduism combines the benevolent and malevolent manifestations of their gods. To the devotees, the gracious, nurturing, and loving mother goddess, Devi, and the demonic, bloodthirsty, and vicious she-devil, Kāli, are one and the same. Yin and Yang, opposites, characterize all creations, even gods. While there is pleasure in creation, there also is pleasure in destruction. However, most devotees find the Absolute more accessible in the benevolent, creative mode than in the fearsome, destructive mode.

Separation and differentiation are characteristics of the dominant hemisphere, while combination and unification are nondominant hemispheric functions. Predictably, creation and experience of art forms are nondominant hemispheric activities. In all probability, the same hemisphere also mediates the closely aligned religious ecstasy.

Like *rasa*, the artistic emotion, religious experiences are also intensely pleasurable. If the experience and expression of art and religion are dependent upon the nondominant hemisphere, the associated pleasure must be mediated by the reward mechanisms of that hemisphere. Indeed, nerve tracts containing dopamine are present in both hemispheres. However, scientists have found that the dominant hemisphere has more dopamine, while the nondominant hemisphere has more norepinephrine. The dopamine in the dominant hemisphere rewards discrete goal-oriented actions (phasic), while the norepinephrine in the nondominant hemisphere provides diffuse background activity (tonic).

Consciousness is that which supports and sustains all mental activities, and neurons containing norepinephrine are abundant in the brain stem. The nondominant lobe that mediates intuitive knowledge and devotion seems to utilize the same neurochemical, norepinephrine, that is linked to consciousness at the brain stem level. Thus, it is conceivable that the more sophisticated, spirituality-enhancing activities, such as intuitive knowledge and devotion, depend upon the nondominant hemisphere that is more evolved than the brain stem.

It is intriguing to ponder the reason art and music should make their appearance so late in the evolutionary chain. The earliest evidence of art in the form of cave drawing dates back about 50,000 years. In the following years, bone carvings, beads, pendants, and so on made their appearance. Art, which started in Europe, became a worldwide phenomenon approximately 30,000 years ago.

Music may have made its appearance as a means of communication, but it does not seem to confer an advantage in survival to the more developed species, notably the mammals, including man. I cannot think of a reason that art and music should continue to evolve and enlarge instead of extinguishing themselves. I am quite confident that evolutionary psychologists, who specialize in coming up with brilliant hypotheses to explicate such conundrums, will rise to the occasion; however, Darwin himself was less certain. In *The Expression of the Emotions in Man and Animals,* he wrote:

> *We can plainly perceive, with some of the lower animals, the males employ their voices to please the females, and that they themselves*

take pleasure in their vocal utterances; but why particular sounds are uttered, and why these give pleasure cannot at present be explained . . . But this leaves unexplained the more subtle and more specific effect of which we call the musical expression of the song—the delight given by its melody, or even by the separate sounds which make up the melody. This is an effect indefinable in language—one which, so far as I am aware, no one has been able to analyze, and which the ingenious speculation of Mr. Herbert Spencer as to the origin of music leaves quite unexplained.[8]

While physicists, mathematicians, and naturalists are awed and overwhelmed by the loveliness of nature, no one has tried to provide a logical reason.

In *On the Origin of Species,* Charles Darwin writes, "Natural selection . . . is a power incessantly ready for action and immeasurably superior to man's efforts, as works of nature are to those of art."[9] Reproductive attraction may be the explanation for the beauty in the case of animals. But the beauty of inanimate objects such as the symmetrical patterns on the wind-swept desert sand, the gentle murmur of the wind, and the majesty of the star-studded night sky defies practical explanation.

The elegance and symmetry of the basic equations that describe the universe, of which the physicists are convinced, are also beyond human logic. In *The Naval Treaty,* Sir Arthur Conan Doyle's stolid sleuth Sherlock Holmes, after gazing at the dainty blend of crimson and green of a moss-rose, observes:

There is nothing in which deduction is so necessary as in religion. It can be built up as an exact science by the reasoner. Our highest assurance of the goodness of Providence seems to me to rest in the flowers. All other things, our powers, our desires, our food, are all really necessary for our existence in the first instance. This rose is an extra. Its smell and its color are an embellishment of life and not a condition of it. It is only goodness which gives extras, and so I say again that we have much to hope from the flowers.[10]

Selfless Self

Good and Evil

Communion with the Divine is central to most faiths. Prayer, rituals, and a lifestyle consonant with the beliefs of a religion assist the aspirant in accomplishing this. Most religions place a great deal of emphasis on doing good and avoiding evil; lengthy lists of desirable and undesirable behaviors are provided. While some of the recommended behaviors are religion-specific, most are nonspecific and shared by all. These universals are of greater interest to us in our quest for underlying brain mechanisms.

After I left India in 1972, I spent three years in England before coming to the United States. Since then, I have traveled widely. In my experience, basic concepts of good and bad cut across caste, creed, religions, and nationalities. A good person is good all over the world, as a bad person is bad.

This raises the serious possibility that basic notions of good and evil are axioms ingrained in the human psyche at the most basic level and not creations of sociocultural beliefs and practices. That is not to say that society has nothing to do with it. Of course, there are culture-specific ones. Muslims accept polygamy, but Christians do not. Christians practice meat eating, but upper caste Hindus do not. Identifying people as untouchable is sanctioned by Hindu orthodoxy, but definitely not by Christians or Muslims. For us to identify relevant brain mechanisms, we have to isolate the universal from the communal.

Some religions clearly spell out what is expected of their followers. The Ten Commandments Moses received from Jehovah are a striking example. Islam spells out the expectations; Islamic law is notable for its clarity and precision. While the Qur'an—the revealed word of God—sets forth broad guidelines, most of the law is based on Shariah, derived from the Hadith, the sayings and teachings of the Prophet as recorded by his friends. Although these teachings are of value and significance, they cannot be easily reduced to a few basic principles.

Theistic Hinduism does not offer much help, either. Part of the problem is that there is no such monolithic religion as Hinduism. There is no single canonical text accepted by all Hindus, no central authority, and no administrative hierarchy.

Of the several ancient Dharmaśāstras (law books), the one written by Manu (Manusmṛti) is generally regarded as the closest to Hindu jurisprudence. Manu's concept of proper behavior was heavily tainted, at least in my opinion, by contemporary concepts of social ordering. The caste system, which served to consolidate and preserve Āryan superiority, undergirds Manu. Manu's concept of duty, primarily but not exclusively, consisted of performance of rites and rituals, discharging social obligations, and staying within the bounds of the caste system.

I do not mean to imply that this is all there is to Manusmṛti; the treatise does address many other moral and ethical issues of interest and practical use. However, many of Manu's concepts of ethics, based on caste hierarchy, can hardly be seen as universal.

This was very much in consonance with the Hindu way of living in India at that time. Rigid ordinances of the caste-based system regimented and regulated every aspect of life; in the process, it strangled and suffocated social freedom, especially for the underprivileged masses.

The major movements in Indian philosophy and religion that took place during this time were revolts against this system. Buddhism falls into the category.

The Buddha, when he set out on his search for a solution for human misery, first resorted to self-mortification. Such an approach had wide acceptance, including Manu's. The Buddha said:

> *I was unclothed, indecent, licking my hand . . . I subsisted on the roots and fruits of the forest, eating only those which fell. I always stood and did not sit . . . I slept on thorns. Dust and dirt of years accumulated on my body. Because I ate so little my limbs became like the knotted joints of withered creepers, my buttocks like a bullock's hoof, my protruding backbone like a string of beads, my gaunt ribs like the crazy raptors of a tumbledown shed. My eyes were sunken deep in their sockets . . . my scalp was shriveled . . . my hair rotted at the roots, fell out if I rubbed it with my hand.*[1]

Buddha, after six years of such rigorous ascetic existence, recognized its futility and abandoned it. The five fellow enlightenment-seekers abandoned the Buddha in disgust as a backslider and one who was weak in his resolve. A pensive Buddha left the cave, in which he'd lived, and walked to a nearby village.

Sujāta, a village girl, provided his first meal of rice pudding. Thenceforth, he avoided the extremes of self-indulgence he had known in his earlier life as a prince and the self-denial he had tried as a recluse. He continued his search for relief and remedy for human suffering. He progressed slowly but steadily in his struggles, and one night under the full moon, in a flood of luminance, the wisdom he was seeking came to him abruptly. As the flimsy veils of unreality fell away, he gazed at the beatific spectacle of truth and reality. He was swept up in a rolling tide of wisdom, serenity, and joy, which remained with him for the rest of his life. He spent the next several weeks immersed in the rapturous ecstasy, recapitulating his thoughts and formulating his plans for the future. After considerable hesitation, he decided to take his newfound wisdom to the suffering multitudes.

Selflessness

The system of philosophy he set forth was simple and straightforward and useful to us in our search for basic concepts. It rested on four pillars known as the four noble truths. The Buddha argued that life was illusory; unlike the Upaniṣads, he saw no redeeming qualities to it at all. Aging, disease, and death, the inevitable consequences of life, made it totally unappealing. It was not just a dream; it was a nightmare.

Selfish attachment to the tinsel and glitter that laced the bad dream was the cause for suffering—the second noble truth proclaimed. In the ever-changing flux of earthly life, everything is impermanent. Loss is inevitable and attachment, therefore, is guaranteed to bring pain.

Yet the situation is not as hopeless as it would seem—the third truth reminds us that there is a way out. To the diehard materialist, abandonment of material possessions and detachment from carnal pleasures might seem to produce a featureless vacuum. In fact, Buddhists themselves call this negative state *śūnyatā,* meaning emptiness; but they consider it highly desirable, in every sense of the word. The third noble truth argues that release from the material world will bring about serenity, peace, and tranquillity hitherto unknown.

The fourth truth delineates the Buddhist way of liberation from trammels of the flesh.

The eightfold path of Buddhism—the Buddhist path to release—consists of right understanding, right thought, right speech, right actions, right livelihood, right effort, right mindfulness, and right concentration. Traditionally, they are appended under three categories known as pillars (*khanda* in Pāli). Right view and right thought fall under wisdom, previously discussed; right speech, right action, and right livelihood are subsumed under conduct; and the remainder fall under meditation. Here we will consider the middle group on conduct.

After the Buddha's death, Buddhism fractured into many fragments and the legitimacy of various doctrines is highly controversial. For our purposes, we will stay with the Dhammapada, the Buddhist treatise which enjoys universal acceptance as the Buddha's own words.

The specific behaviors Buddhism bans are killing, stealing, adultery, lying, and intoxication. Buddhism believes that these undesirable behaviors can be abolished by what it calls purification of the mind, which, by and large, is expurgation of egoistic thoughts, behaviors, and deeds. Buddha said, "Not by rituals and resolutions, not by much learning, nor by celibacy, nor even by meditation can you find the supreme, immortal joy of *nirvāṇa* until you have extinguished your self-will" (The Dhammapada, 19, 271–272).

The Buddha argued consistently and strongly that empirical self is simply a figment of one's imagination and that, in reality, there is no such thing. In addition, he firmly believed that identification with the nonexistent empirical self was the root cause for human suffering. Deliverance is really and truly release from the empirical self. "Him I call a holy man who has turned his back upon himself. Homeless, he is ever at home; egoless, he is ever full" (The Dhammapada, 26, 415). "Him I call a holy man who is free from I, me, and mine, who knows the rise and fall of life. He is awake: he will not fall asleep again" (The Dhammapada, 26, 419).

As the outer walls of the ego melt and run, self merges with nonself. Pain to others will begin to hurt more than pain to oneself, and giving becomes more pleasurable than taking. "One who injures living creatures is not noble. Those who hurt no one are noble" (The Dhammapada, 19, 270). Others' happiness becomes more joyous than one's own. Self enlarges and expands to include everything living and nonliving. The enlightened strive to avoid insult and injury to the environment.

The Buddha believed that it is possible to live our lives out without exploiting, deforming, and disfiguring nature. "The wise live without injuring nature, as the bee drinks honey without harming the flower" (The Dhammapada, 4, 49). This state of mind where the empirical self does not exist and one feels totally related to, associated with, and absorbed in a collective whole is called *dharma* (*dhamma* in Pāli). "He is a true sage who follows the *dhamma*, meditates on the *dhamma*, rejoices in the *dhamma*, and therefore never falls away from the *dhamma*" (The Dhammapada, 25, 364), The *dharma* is usually represented symbolically by a wheel. The spokes of the wheel are the rules of pure conduct; justice is the uniformity of their length; wisdom is the tire; modesty and thoughtfulness are the hub in which the immovable axle of truth is fixed.

Identification with the *dhamma* is not possible so long as one is entrapped in the narrow concept of self. According to the Buddha, "Self is a fever; self is a transient vision, a dream, but truth is wholesome, truth is sublime, truth is everlasting."

Anxiety and fear are both related to self-preservation and both have the empirical self as its basis. Destruction of ego will lead to the dissolution of both anxiety and fear. "Selfish desires give rise to anxiety; selfish desires give rise to fear. Be unselfish, and you will be free from anxiety and fear" (The Dhammapada, 16, 215). The selfless possesses nothing and has nothing to lose. Death is fearful only so long as you are attached to life.

The Buddha impugned the notion of self, its roots, and phenomenal reality upon which it was based. He vehemently denied the existence even of a personal soul.

Nirvāṇa, the Buddhist term for deliverance and enlightenment, literally means "that which is extinguished." When the concept of an empirical self is extinguished, what remains is a state of wakefulness, sustained and supported by *dharma*. The natural tendency, the Buddha reminds us, is to be enthralled by the empirical ego and to pursue activities that nurture and aggrandize it. If you long for liberation, the Buddha said, you have to swim against the current.

The Buddha transformed Hinduism and Indian philosophical thought forever. Buddhism loomed large and eclipsed theistic Hinduism, which remained in its shadow for the next thousand years or so. The Buddha was a cataclysmal earthquake that shook down the existing totems of priesthood and ritualism. He was a torrential monsoon shower that washed away the dust and dirt from religious thought in India. He was the first coming of spring, fresh, beautiful, and unbelievably soothing.

Some Upaniṣads also express views similar to the Buddhist doctrine of selflessness. According to Iśa Upaniṣad, "He who sees all beings in his own self and his own self in all beings, he does not feel any revulsion by reason of such a view"(6). However, most Vedic literature is wrapped up in lofty philosophizing, with little said about day-to-day life. The Upaniṣadic ideal is a recluse and not a person of action. Gītā, introduced earlier, departs from this traditional ideal.

It repudiates inaction in no uncertain terms: "Action is greater than inaction: perform thy task in life, even the life of the body could not be sustained if there were no action" (3, 8). However, the concepts of action and inaction Gītā espouses do not correspond to commonly held notions. Meditation, for example, is not inactivity at all; in fact, it is the most intense activity. Inactivity is laziness, sloth, and indolence. Action, Gītā reminds us, must not be selfish; in fact, selfish desire is identified as the root cause for all evil. "It is greedy desire and wrath, born of passion, the great evil, the sum of destruction: this is the enemy of the soul . . . wisdom is clouded by desire, the ever-present enemy of the wise, desire in its innumerable forms, which like a fire, cannot find satisfaction" (3, 37, 39).

Gītā, basically theistic, instructs the devotees to dedicate their work to the Divine and not to personal aggrandizement: "Offer all thy works to God, throw off selfish bonds, and do thy work. No sin can then stain thee, even as waters do not stain the leaf of the lotus" (5, 10). According to Gandhi, the central message of Gītā is action unmotivated by personal gain *(nishkāma karma)*.

Gītā equated the narrowly defined empirical self with man's lower nature (18, 25). Śāṁkara behooves us to give up such notions as "I am my body, my intellect and my ego" (Vivekacūḍāmaṇi, verse 296). He warns that the empirical self can never be satisfied; the more one acquires, the more one will desire (Vivekacūḍāmaṇi, verse 313). In the manner of the Buddha, he too reminds us that erasing the empirical self is no easy task as it tends to come back, as rain clouds gather together after the wind disperses them (Vivekacūḍāmaṇi, verse 309) and moss springs back to cover the water surface after it is swept aside (Vivekacūḍāmaṇi, verse 324). One has to work on being selfless, diligently and assiduously.

Christianity, like Buddhism and Gītā, places tremendous importance on the virtue of selflessness. As expressed in the Gospel according to Luke, "For whosoever exalteth himself shall be abased; and he that humbleth himself shall be exalted" (14.11). In the Gospel according to Mark: "Whosoever will come after me, let him deny himself, and take up his cross and follow me. For whosoever will save his life shall lose it; but whosoever shall lose his life for my sake and the gospels, the same shall save it" (8.34, 35). The Sermon on the Mount also underscores the importance of humility, selflessness, kindness, purity, simplicity, peacefulness, and righteousness (Matthew, 5.3–11).

Although we are today by and large self-centered, we are still attracted to unselfish people and things. We love small children in whom intellectual sophistication and egoism is conspicuous by its absence. Children with immature egos and empirical selves display the purity and serenity of the unbound Absolute. Yoga and the associated bliss are more accessible through them. Once Jesus called a little child to him and set him in the midst of his disciples. He told them, "Whosoever shall humble himself as this little child, the same is greatest in the kingdom of heaven" (Matthew, 18.1–4). Parallel passages can be found in the Upaniṣads. According to Subālā Upaniṣad "One should cultivate the characteristics of a child, which are non-attachment and innocence" (13, 13). For similar reasons, most people find associations with pets blissful.

We have been able to identify some universals common to Vedic Hinduism, Buddhism, and Christianity, if not to all religions. All of them instruct us to abandon the narrow concept of empirical self for a larger, inclusive self; care and concern for, and sensitivity toward others will follow. I do not mean to imply that these are the only universals or that these represent the three great religions. There are any number of principles specific to each that differentiate them. But for our purposes, we have to stay with what is common to all.

The Contemporary Relevance of Selflessness

There are a few issues that need to be addressed before attempting to translate these phenomena into brain mechanisms. First of all, these concepts are based upon religions founded close to 2,000 years ago. It is unclear to what extent they can be extrapolated to present times. One would have to make sure that these are not lofty idealisms, impractical and impossible in the present day. There is no point in searching for brain mechanisms unless they are relevant to life as we know and understand it.

Selfless deeds aimed at the establishment, restoration, and preservation of righteousness in the world with no importance given to personal gains and losses constitute Karma yoga, or union through action. Many people familiar with Karma yoga in India and elsewhere regard Mohandas Gandhi as a prime example. There are any number of people all over the world who have lived and still live by these ideals. When I choose Mohandas Gandhi as an example, I do not mean to imply that he is the only one or the best one. I chose him because he was well-known as a result of his involvement in politics, and the effectiveness of his methods was demonstrated by the results he accomplished. There is ample documentation of his thoughts and actions, and lastly, he lived in the twentieth century, and there are many people alive today who knew the living Gandhi.

Gandhi was consumed by his desire for yoga. He wrote in his autobiography, titled *The Story of My Experiments with Truth,* "If I found myself entirely absorbed in the service of the community, the reason behind it was my desire for self-realization. I had made the religion of service my own, as I felt that God could be realized only through service. . . What I want to achieve, — what I have been striving and pining to achieve these thirty years, — is self-realization, to see God face to face, to attain Mokṣa. I live and move and have my being in pursuit of this goal."[2]

Gandhi's philosophy of life was relatively simple. It had two essential components; namely, one of destroying the concept of self, and secondly, embracing the universal principle of truth (*dhamma* in Buddhism). In his book *Letters to a Disciple,* he wrote: "If I succeed in emptying myself utterly, God will possess me. Then I know that everything will come true, but this is a serious question when I shall have reduced myself to zero. Think of 'I' and 'zero' in juxtaposition and you have the whole problem of life in two signs."[3]

Gandhi attempted to do so by decreasing egoistic indulgences and aspirations in all spheres. He wore very simple garments and always preferred to travel in third class. In *The Story of My Experiments with Truth,* he wrote: "The seeker after truth should be humbler than dust. The world crushes the dust under its feet, but the seeker after truth should so humble himself that even the dust could crush him. Only then, not until then, will he have a glimpse of truth . . . Christianity and Islam also amply bear this out." Gandhi considered self-purification as absolutely essential; it basically meant selflessness and freedom from the trammels of the flesh.

According to Gandhi: "God can never be realized by one who is not pure of heart. Self-purification, therefore, must mean purification in all walks of life . . . To attain perfect purity one has to become absolutely passion-free in thought, speech, and action; to rise above the opposing currents of love and hate, hatred, attachment, and repulsion."[4]

Attraction and aversion arise from the dichotomous phenomenal world. Both Buddhism and the Upaniṣads talked about a superior existence above and beyond these. God, Gandhi believed, can be accessed by embracing truth. The method Gandhi adopted in fighting the British colonization of India, which he regarded as unjust *(adharma)*, was *satya agraha.* Gandhi himself explains the term in his autobiography; *satya* stands for truth and *agraha* for firmness. Together they mean "firm establishment in truth."

The word "truth" had a very specific meaning for Gandhi. "But for me, truth is the sovereign principle which includes numerous other principals. This truth is not only truthfulness in word, but truthfulness in thought also, and not only the relative truth of our comprehension, but the Absolute Truth, the Eternal Principal, that is God."

Dr. Martin Luther King Jr., another votary of truth, espoused Gandhi's philosophy and principles. Dr. King said about Gandhi: "If humanity is to progress, Gandhi is inescapable: he lived, thought, and acted inspired by the vision of humanity evolving toward a world of peace and harmony. We may ignore him at our own risk."[5] The universals I mentioned above can be found in the "Ten Commandments" that the volunteers who followed Martin Luther King's nonviolent movement abided by:

> *I hereby pledge myself, my person and body—to the non-violent movement. Therefore, I will keep the following Ten Commandments: 1) Meditate daily on the teachings and life of Jesus.*

2) Remember always that the non-violent movement in Birmingham seeks justice and reconciliation—not victory. 3) Walk and talk in the manner of love, for God is love. 4) Pray daily to be used by God in order that all men might be free. 5) Sacrifice personal wishes in order that all men might be free. 6) Observe with both friend and foe the ordinary rules of courtesy. 7) Seek to perform regular service for others and for the world. 8) Refrain from the violence of fist, tongue, or heart. 9) Strive to be in good spiritual and bodily health. 10) Follow the direction of the movement and of the captain.

Pleasure, Self, and the Brain

Discussions of brain mechanisms relevant to these principles would seem bizarre and far-fetched. Yet we have to remind ourselves that Gandhi and King were mortals with brains. The thoughts and experiences that guided them were brought about by mechanisms that were present in the minds and brains of many who went before them. While such experiences cannot be reproduced in the laboratory for investigation, it is possible to speculate on the related neurophysiological mechanisms.

The first question is why anybody would want to be good. All activity is pleasure-directed. Such simple functions as feeding, reproducing, and eliminating, as well as more complex ones such as making money, getting promotions, and becoming famous, are all pleasurable. In fact, living organisms do not engage in unrewarded activity. But Gītā reminds us that reward does not have to be in the result—it can be inherent in the action itself. "Even as the unwise work selfishly in the bondage of selfish works, let the wise man work unselfishly for the good of all the world" (3, 25).

There is reward in being honest, kind, and self-sacrificing. All over the world, people who lived and died for the sake of others are honored and remembered. Jesus and the Buddha gained very little in a material sense, neither did Martin Luther King, David Livingston, or Mother Teresa.

In the Yin-Yang world, pleasure is closely coupled with pain; both are sides of a single coin. Mundane pleasure has no independent existence; it is always relative to pain. Therefore, pleasure of this type may be described in terms of avoidance of pain. Typically, mundane pleasure is associated with gains and pain with losses. Indian thinkers mention *ānanda*, another type of pleasure with no contrasts and independent of gains and losses. The peace and tranquillity associated with art, dreamless sleep, and spiritual experiences resemble this.

Pleasure occurs when certain neurotransmitters, which facilitate signal transfer between neurons, are released. Norepinephrine and dopamine are two of the reward chemicals identified in the brain. Norepinephrine and dopamine can be found in both hemispheres. According to psychologists Don Tucker and Peter Williamson, there is a higher concentration of dopamine in the dominant left hemisphere (in predominantly right-handed people) and a higher concentration of norepinephrine in the nondominant right hemisphere. Dopamine in the dominant hemisphere rewards phasic activities, while norepinephrine in the nondominant hemisphere rewards tonic events. Phasic activities are specific, goal-directed behavior, whereas tonic activities are more diffuse and less focused. The dopamine-laced dominant hemisphere mediates activities on the basis of the linear system of operation of that hemisphere. It is more proficient at the use of logic and reason in targeting well-defined goals most likely to benefit the organism. It is also likely to have a concept of self sharply demarcated from non-self. The dominant hemisphere is particularly adept at separation and divisions; it helps us decide how to invest well, how to get ahead of competition, and how to succeed. Some scientists believe that the empirical concept of self is language based. As we have seen, language is a dominant hemispheric function. The dominant hemispheric activities are rewarded by the release of dopamine, present in greater abundance in that hemisphere. The nondominant right hemisphere, on the other hand, is holistic in its style of operation. It is less efficient in separating self from surroundings and, therefore, may not cater to the needs of empirical self as efficiently as the other hemisphere. It is also less skilled in the use of logic and reason. It may navigate us to a panoramic view, persuade us to listen to a song, or prompt us to play with the family dog, none of which may benefit us in a material sense. All of these activities are holistic, involving dissolution of boundaries, and the associated pleasure is diffuse but of a distinct quality. Unlike the phasic activities the dominant hemisphere specializes in, the tonic activities of the nondominant hemisphere are less formed and precise.

The dominant hemisphere gets sensory information primarily from the opposite side. The nondominant one, on the other hand, receives sensory input from both sides and integrates it. Access to the full range of sensory reports gives the nondominant hemisphere superiority in creating the appropriate tone and temper. Norepinephrine, which the nondominant hemisphere has in greater abundance, as we have seen, is also intimately involved in the arousal system housed in the brain stem. Norepinephrine-containing fibers originate from the brain stem nuclei and locus ceruleus and innervate the entire brain. The recent claim of denser norepinephrine innervation of the nondominant hemisphere would indicate that this hemisphere has closer connections with the arousal system. Arousal is the physiologic basis for empirical consciousness.

Pleasure motivates activities involving both hemispheres. Such left-hemispheric activities as winning a debate, getting promotions at work, or making money are rewarded by a flow of pleasure chemicals in that hemisphere. Similarly, right-hemispheric activities such as being with family and listening to music also give pleasure. While it is reasonably clear that the dominant-hemispheric activities listed above stimulate dopamine release, it is less clear whether the nondominant activities are associated with norepinephrine release. Dopamine is also present, albeit in smaller quantities, in the nondominant hemisphere. It is quite conceivable that the nondominant, pleasure-producing activities are associated with dopamine only, norepinephrine only, or a combination of the two. The main point here is that the quality of pleasure associated with the dominant hemispheric activity is distinctly different from that associated with a nondominant hemispheric action. Unfortunately, the differences cannot be articulated. Perhaps part of the reason is that language, upon which articulation depends, is predominantly a dominant-hemispheric activity, and it may not be as proficient in expressing a mood state associated with activity in the nondominant hemisphere.

Drugs like LSD, mescaline, and marijuana linked to religious ceremonies have been found to activate the nondominant hemisphere. Art and music, also associated with worship, stimulate that hemisphere. Being with family, enjoying scenery, or playing with a pet may also involve the same hemisphere. The above-mentioned drugs and activities produce a unique type of pleasure that may be norepinephrine-mediated. The nondominant type of well-being may be closer to the bliss, or *ānanda,* of consciousness that may also be mediated by norepinephrine. The reinforcer for nondominant hemispheric activities may indeed be norepinephrine. It may be this neurochemical that persuades us to be kind, sympathetic, and giving even when it bodes the individual no good in a material sense.

Sensory information that streams into the brain through the sensory channels is sifted for pleasure and pain cues by a brain structure called the amygdala, buried deep in the temporal lobe. Attention is selectively focused on sensory cues thus flagged. The organism avoids information likely to induce pain and selectively focuses on cues that herald pleasure. While the left amygdala may go for activities that benefit the empirical self, the amygdala of the nondominant side may pick holistic cues that promise a different type of reward.

The mundane world is impermanent; there are no permanent owners here. We are all passing through, and any pleasure linked to ownership is guaranteed to cause pain in the long run. Spiritual teachings herald that worldly pleasures bring pain in the long run and that they are inferior to spiritual pleasures. Listening to a song, enjoying the idyllic qualities of nature, and feeding wild birds do not involve ownership. There is less risk of loss. Actions designed to benefit a self temporally and spatially isolated will bring pain in the long run, since a self so conceived will decay and die.

From an evolutionary perspective, the dominant hemispheric type of reward is easier to explain. According to Charles Darwin, organisms endowed with an advantage over others survive the competition for resources necessary for sustenance. Thus, survival depends upon superiority of one organism over another. Therefore, reinforcement of behavior that confers superiority makes a great deal of sense.

In the animal kingdom, strength in general is the law. Even among chimpanzees, the sick and the weak are neglected. However, the skeletal remains of a 60,000-year-old Neanderthal man indicated that he was cared for in spite of his severe incapacitation from birth injuries. This is clearly related to a sense of community, which evolved as the brain became larger and more complex. Religion also made its appearance around this time, as evidenced by burial sites and the placing of animal carcasses over the graves.

Belonging to a group insured survival for early humans, and reinforcement of group-related activities in general can thus be explained. It is unclear, from the evolutionary perspective, why listening to music, being kind to other people, and having concern for the environment should be rewarded at all.

Even more confusing is the contemporary emphasis on equality and justice for all, preservation of the past, and protection of the environment. Such attitudes seem to contravene basic tenets of evolution and survival of the fittest. Indeed, scientific publications are available on the deleterious effects of disregarding the time-tested principles of evolution. If these tendencies that are unrelated to self-preservation made their appearance through chance mutations, it is surprising that the inexorable force of selection did not snuff them out.

Aldolph Hitler was dedicated to survival of the fittest and did his best to improve the quality of the gene pool by eliminating those who carried disadvantageous genes. He actively and energetically worked for the preservation and betterment of a narrowly defined concept of self. From a logical point of view, what Hitler attempted to do makes perfect sense. Yet he is demonized and looked down upon by most of the "civilized" world.

Being honest, kind, and righteous often portends no good for the individual in question. It is debatable whether honesty and kindness are the best tools in the race to get ahead. One is often caught in the dilemma between the goal and the means to the goal.

While success is highly desired and admired, we are not sure how important the means to success is. Lying, cheating, and exploiting are often shortcuts to success. However, such behavior does not benefit society, mankind, our planet, and the universe. The deciding factor would seem to be the concept of self: a narrowly defined egoistic one or a larger inclusive one.

Evolutionary biologists and psychologists may explain the appearance of humanitarian values on the basis of their survival value for the community as a whole. It is conceivable that people have become more concerned about others because they realize that that is the logical thing to do. Man is a social animal and unhappiness and disgruntlement in any sector will eventually destabilize and corrode the establishment and ultimately destroy the individual. Although there is some substance to this argument, I believe that most people are concerned about others because they feel that is the right thing to do. I do not think it is based upon logic. Nevertheless, it is a debatable point.

Scientists have argued that science is by and large responsible for human progress. They warn us that resurgence of religion may sink us back into the murky depths of superstition and ignorance. The whole issue revolves around what one means by progress. If civilization means technological advancement, improved quality of living, and longevity, they may be correct. On the other hand, if civilization is compassion, kindness, and a consuming desire for fair treatment, they are not. While technological progress has increased the comfort of our existence and delayed the onslaught of death, it has also provided us with weapons of mass destruction of ever-increasing power. According to Dr. Martin Luther King Jr., knowledge in the hands of people without character is dangerous.

The drugs and activities that enhance nondominant function also bring about a lessened, if not lost, sense of self. Temporal and spatial lines of demarcation become thinner and disappear. Scientific evidence supports the notion that the nondominant hemisphere, which receives bilateral sensory input and integrates it into a collective whole, has an expanded notion of self. Thus, there would appear to be both linear and holistic models of self (Figure 8.1).

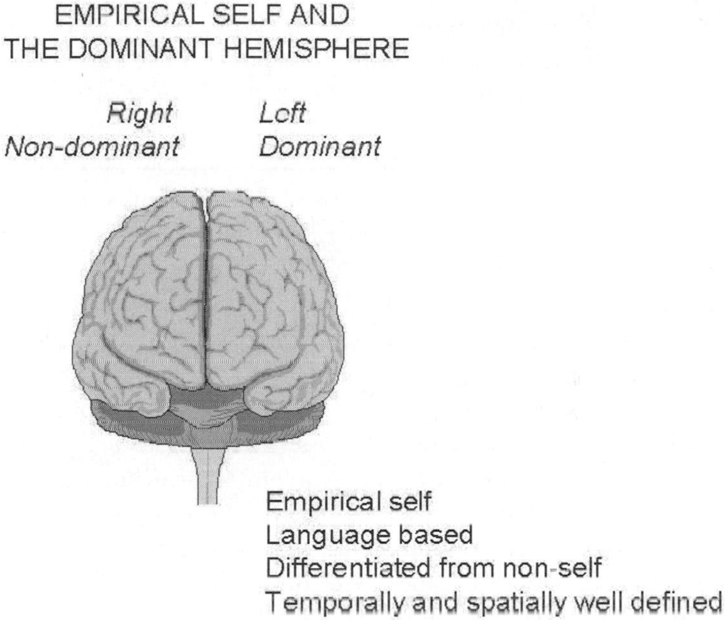

FIGURE 8.1 *The language-based empirical self that is sharply differentiated from the surroundings is probably mediated by the dominant left hemisphere.*

The ability to manipulate the environment, which is key to the protection and preservation of life, is dependent upon the ability to reason. The ability to communicate needs and wants is also central to survival. In right-handed people, the left hemisphere, which mediates language and the logical approach, is predictably dominant. It overrides the other hemisphere in most activities. The dominant hemisphere would therefore seem to represent mundane self. However, the nondominant hemisphere also mediates self-awareness, but probably in a different way. The vague and diffuse sense of self-awareness associated with the right hemisphere is harder to define.

The technique devised by J. Wada in 1949 for the determination of the hemisphere responsible for speech also allowed for evaluation of the hemispheric contribution to consciousness. The technique involved injection of a barbiturate into the carotid artery that perfuses the hemisphere of interest (chapter 7, figure 4). Immediately following the injection, the hemisphere would be selectively anesthetized for a short while, with temporary loss of the functions it mediated. The technique was most useful in identifying the hemispheric dominance for language. The technique, however, provided other information as well, some of considerable heuristic significance.

For example, several investigators reported more frequent loss of consciousness following anesthetization of the dominant hemisphere. This seemed to suggest that the dominant hemisphere mediated ordinary waking consciousness. Inactivation of the dominant hemisphere is unlikely to cause total unconsciousness, since its surgical removal does not cause it. None of the investigators addressed the obvious question about the quality of the remaining nondominant hemisphere–mediated consciousness.

Part of the problem is that dominant hemispheric inactivity is associated with impaired ability to speak and write, which would significantly limit the subject's ability to communicate the qualitative aspects of the conscious experience. Lesions of the dominant hemisphere have also been found to cause impaired consciousness more frequently. The effects of drugs that selectively activate the right hemisphere support the possibility that the consciousness mediated by the right hemisphere is qualitatively distinct. A few other experiments have been carried out on the hemispheric contributions to the concept of self. In general, the dominant left hemisphere was found to be responsible for self as distinct from non-self. Patients with left hemispheric damage had more severe disruption of their self- concept.

Our experiments have indicated that THC and marijuana activate the nondominant hemisphere. Other investigators have reported similar nondominant hemispheric effects for LSD and mescaline. These drugs produce selective alterations in the quality of conscious experience. However, though substantial and profound, it is difficult to put the experience into words. The effects of these drugs on the mind were reviewed in the previous chapter. In summary, it may be stated that these drugs blur the boundaries of self, both temporally and spatially, and cause consciousness to become less sharply defined and to assume a surreal hue. There is often a softening of the perception-based concept of reality.

We have been self-aware from our unicellular days to the present; both the amoeba and man are aware of themselves. The chimpanzee can identify its mirror image. Comparisons, in a qualitative sense, are not possible, for it is difficult to find out much about human self-awareness, not to mention that of an amoeba. However, as the nervous system originated and evolved, the concept of self became more extensive and complex. The cerebral hemispheres, the latest evolutionary step, brought further elaboration and extension of self-awareness.

Dominance of the left hemisphere left its impression on "I," the linguistic concept of self, as spatially and temporally distinct from the surroundings. Consciousness, thus encapsulated and cocooned, corresponded to the concept of egoistic self.

It is conceivable that the destruction of empirical self that the founders of major religions talk about involves debunking the linear dominant hemispheric model in favor of the holistic, nondominant one. Kindness, sympathy, fairness, and honesty that benefit the society and environment do not confer advantages for the empirical self. One seldom gets rich and powerful by being generous and giving. These traits support and sustain the inclusive model of self that the nondominant hemisphere generates.

I have said a great deal about hemispheric functional asymmetries, and a few disclaimers would seem to be in order. Linear and holistic modes of operation for the dominant and nondominant hemispheres should not be seen as black and white. Any given activity, whether it is language, logic, or music, can be taken apart into linear and holistic components. Thus, it is not correct to say that music is exclusively nondominant while language belongs primarily to the dominant. Most activities, cognitive as well as behavioral, stimulate components of both hemispheres.

Attempts have been made to differentiate personalities into left and right hemispheric. Such concepts have found little acceptance in scientific circles. While the left and right hemispheres may have developed certain functions strongly, at a basic level, both hemispheres are able to perform similar tasks. It is more correct to say that the brain possesses two modes of information processing, linear and holistic, and that in right-handed people the former is better seen in the left hemisphere and the latter in the right hemisphere. I may add that this is an area of considerable controversy, and firm conclusions will have to be postponed.

Extremes of the two orientations are rare; most of us have some of each. The nondominant hemisphere and the dominant one evolved from the same source; therefore, it would be incorrect to maintain that all dominant-hemispheric activity is nonspiritual. For example, language, a dominant-hemispheric activity, has been ascribed great spiritual significance.

Many Indian grammarians, notably Bhartṛhari, who lived during the sixth century, deified language and the formation of language. This tendency, however, can be found even in the much earlier Ṛg Veda that deifies "vāc," or "word."

The Upaniṣads accepted and acknowledged this line of thought. According to Bṛhadāranyaka Upaniṣad, "Speech is Brahman, and this is lord" (I, 3, 21). Bhartṛhari's statement from his Vākyapadīya, "The eternal one, the imperishable Brahman of which the essential nature is word," is reminiscent of the first sentence of the gospel according to John, "In the beginning was the Word, and the Word was with God and the Word was God" (1.1). To Bhartṛhari, the articulated word is different from the impulse that precedes it; the impulse to create is closely aligned with consciousness. *Śabdabrahman,* or undifferentiated speech, transforms thorough the mechanism of "*spoḍa*" into *nādabrahman*, or speech. Language that is dependent upon time is a created element, while *śabdabrahman* is not.

Bhartṛhari saw grammar as the sacred instrument that directed and defined the transformation of the abstract into the manifest. He elevated grammar *(vyākaraṇam)* as the highest austerity *(tapas)* when in Vākyapadīya he uttered, "Tāpasam uttamam, tapah vyākaraṇam." His concept of grammar was somewhat similar to that of the physicists' view of the fundamental laws of physics that describe the universe. Language given form and foundation by grammar, and after embellishment by the poet's talent, becomes as exquisite an art form as any.

Christians share Bhartṛhari's belief in the close ties between language, consciousness, and divinity. The second chapter of Acts describes speaking different languages as a manifestation of the Holy Spirit. "And they were all filled with the Holy Ghost and began to talk in other tongues, as the Spirit gave them the power of utterance" (Acts, 2.4). In the tenth chapter of Acts, Peter described even the Gentiles speaking "with tongues" to magnify God as the Holy Spirit descended upon them (Acts, 10.46). In contemporary Christianity in the West, many churches, certainly the most charismatic ones, accept tongue speaking as a gift of the Holy Spirit. There is nothing inherently unholy about the dominant hemisphere; one can make any activity, dominant or nondominant, unholy by making it serve selfish goals.

It would seem that the narrowly defined concept of self, which was strong among animals, is losing its hold on humans. Self has expanded to include others who differ from us in gender, religion, and race. We have become more sensitive to the beauty that surrounds us in our neighborhoods, our planet, and the galaxy. We are reaching out, looking for life forms on other planets, so that we can tell them about us and learn about them. It would appear that, for no clear reason, the rolling tide of evolution is carrying us forward in a process of yoga. Yoga, as we have seen, is blissful.

We seem to have identified a new pleasure, one that is different from that associated with selfish gains. The new pleasure, which is considered more civilized, is associated with caring for others more than for self, with justice for all and not for a few, and with giving and not taking.

According to the great Buddha, "If a man who enjoys a lesser happiness beholds a greater one, let him leave aside the lesser to gain the greater" (The Dhammapada, 21, 290).

Getting Past Altered Consciousness

Meditation

Swollen rivers indicate heavy rains in the mountains, heavy winds portend an approaching thunderstorm, and the vesper songs of birds herald the sunset. The beatific experiences of the yogas suggest a deeper wellspring. The yogas we have considered thus far provide hints about the Absolute. Profound though these experiences can be, they are still oblique and tangential. We rely on the shadows to know the sun. We are told that we can intuit the divine. We sing praises to the unseen God's benevolence and magnanimity. We infer righteousness is God's way. Using clues from the peacock we speculate about the egg. But if we desire definite information about the peacock egg, we will have to study it directly.

If the uncreated Absolute lies within the labyrinth of our minds, there should be a way to access it without having to circumambulate. The Upaniṣads declare that the central core of the subjective self is the transcendental Absolute and that it creates the rest. If such is the case, we should be able to reach the Absolute by directing our attention inward into the depths of our being. This is the royal path, Rāja yoga, which includes meditation.

I am uneasy with the confusing, polyvalent term "meditation." In its present usage, at least in the United States, it is only peripherally related to Indian philosophy and yogas. The Sanskrit word *dhyāna,* which is translated as meditation, has strong spiritual and religious connotations that meditation may or may not have. In its secularized version, meditation is at least as often a relaxation treatment for psychiatric and psychosomatic disorders as a spiritual exercise. It is as often, if not more often, practiced in gymnasiums than in places of worship. Ideally, *dhyāna* should mean contemplation or reverie and not relaxation. I will use the term "meditation" in that sense.

Meditation would seem to be as old as human civilization itself. Indus Valley terra cotta seals from more than 3,500 years ago show meditation. The Vedic Āryans practiced a pantheistic religion. They worshiped a multitude of gods and goddesses, many associated with such natural phenomena as dawn and dusk, sun and moon, and fire and wind. As time progressed, the nontheistic spiritual element became more prominent. While some took to the faction with a theistic bent that emphasized rites and rituals *(karma kāṇḍa),* others sought spiritual experience through intuitive knowledge *(jnāna kāṇḍa).*

The Upaniṣadic sages in pursuit of such spiritual experiences withdrew from worldly life to their sylvan abodes. Meditative practices mixed in with devotionalism gave rise to the ancient concept of *tapas. Tapa* literally means "inner heat," and *tapas* represents practices designed to fan the inner "flames." Single-mindedly, they sought to enkindle the divine spark within and cause it to enlarge and engulf the entire mind and body.

Contemplation was not unique to India. In fact, it was common to all faiths and religions, and often it was mixed in with prayer. There are close similarities between Rāja yoga and prayer. We search for God everywhere, and since mind is the mirror where we see and experience everything, our search cannot be external to the mind. Those who have found God have done so only with and within their minds. Prayer calls for attention focused on a mental locus. Devotees who seek the divine in objective space need to be reminded that the objective space is a component of the mind. Even those who believe God is in the sky do not live in roofless buildings in their search for God. The search is always internal, at least in a neurophysiological sense. Places of worship are cloisters where the devotees are shut off; all perceptual experiences are considered distractions, and total silence is essential. Although songs and sacraments are conducive, silence and solitude take us to the peak experience of contact.

As practiced by the desert monks around the fourth century, Christian meditation had Jesus and verses from the Bible as objects of contemplation. Sufis, the Muslim mystics, direct their meditative efforts toward Allah. Theistic Hindus focus on their beloved gods or goddesses. The major and perhaps the only difference between Rāja yoga and prayer is this exclusive focus on a godhead in the latter. In devotion, the search is for the godhead; when the god is devoid of name and form it becomes meditation.

Silent entreaty for communion with the divine was central to the Jewish and early Christian faiths. There are many references to meditation in both the Old and New Testaments. In Kabbala, the ancient spiritual tradition of the Jews, meditation was used as a vehicle to higher states of mind. The Bible also recognizes meditation and fasting as important spiritual practices (Matthew, 6.6, 17–18). Meditation was fundamental to early Christian monastic life. Monasteries were situated away from human habitation, on top of unscalable sheer cliffs or in the heart of arid, uninhabitable deserts. Scriptural knowledge in its own right was inadequate; each monk had to engage in a rigorous personal search for divine communion. Simplicity, austerity, and self-denial were central to the search.

A number of Christian scholars, notably St. Augustine, advocated meditative practices similar to those recommended by ancient Indians. The Christian mystics, John of the Cross and Teresa of Avila, reached states of spiritual ecstasy identical to those described by Eastern, non-Christian spiritual masters. In the contemporary, secularized version of Christianity, such beliefs and practices seem to have fallen into disfavor.

Contemplation bereft of theism is most clearly represented in early Buddhism. While Hindus contemplate Brahman, a nameless and formless entity, the Buddhists focus on emptiness. In some ways the Buddha's thinking resembles that of contemporary neuroscientists, especially behaviorists.

First and foremost, the Buddha strove to base his entire system of belief on logic; there was no room for mysticism. He provided explicit descriptions of the path to liberation he found. It was open and available to all. The recommended course of action was loud and clear and results totally predictable. Thus, the Buddhist descriptions are especially useful to us in our pursuit of relevant brain mechanisms. The Buddha was totally indifferent toward the existence or nonexistence of a god. Release, so far as he was concerned, was entirely dependent upon what we do or do not do. The meditation technique he recommended was very organized, systematic, and precise.

As was discussed previously, the Buddha first tried severe self-mortification as recommended by the Vedas and Manu, (chapters 2, 3 and 9) but he rejected it. His "*madhyāmaka pratipad,*" or "middle path," avoided both extremes and involved meditation sustained and supported by a system of applied philosophy. He wanted his religion to be simple, easy to understand, and accessible to all. Although we are primarily interested in meditation here, ideally it should not be separated from the body of teachings the Buddha imparted. Meditation in its own right is meaningless. For meditation to bring the desired results, it has to be couched in the necessary changes in one's worldview and lifestyle. The reader should consult one of the many excellent publications on Buddhism to fully appreciate the topic.

There were similar movements within Vedic Hinduism, before and after the Buddha. The nontheistic Upaniṣads also refer to yogic practice specifically and in detail. The Buddha learned techniques of yoga and meditation from well-established experts of the time. The exact time, place, and authorship of the original technique of yoga are in dispute. The yoga *sūtra* of Patañjali, the oldest and best-known available textbook on the subject, has been assigned to the second century B.C., although some authors feel it could have been written as late as the fourth century. We will consider the Patañjali approach and an abbreviated version of the Buddhist technique before considering relevant neurophysiology.

Preparation for Meditation

Patañjali's yoga *sūtra* is believed to be an offshoot of the Sāmkhya philosophy discussed earlier. Samkhya recognizes two principles. Prakṛti is the lifeless entity of which the manifested world is derived. Puruṣa, the second principle, is consciousness that endows Prakṛti with life. The yoga *sūtra* enables the aspirant to cut through the outer husk of Prakṛti to access Puruṣa. The theory holds that the central reality of self can be reached only by suppression of mental activities. According to Patañjali, the mind in its normal state, is restless, with multiple fluctuations. It has to be stilled in order for us to reach its inner depths. In yoga *sūtra,* Patañjali defines yoga as the technique to stop mental fluctuations. In this chapter (and in Indian literature in general), the term yoga usually means Rāja yoga or Patañjali yoga, unless otherwise specified.

Like the Buddhists, yoga *sūtra* also insists that enlightenment is not possible without a radical lifestyle change. Many present-day yoga practitioners and teachers unfortunately leave this out. Yoga is widely used both in India and elsewhere for a number of wide-ranging purposes including stress management and pain relief. Some engage in it for a sense of physical well-being, while others have no clear-cut purpose. Although there is absolutely nothing wrong with this shotgun approach, it should be kept in mind that yoga was developed as an exercise to attain *samādhi,* or union with the Absolute. Similarly, the Buddhist meditation techniques had *nirvāṇa* as their goal. The difference between yoga with a spiritual goal and without one is similar to that between a wine-tasting ceremony and receiving the Eucharist. Both involve wine drinking, but there exists a world of difference between the two in purpose and preparation.

Normally, we perceive manifestations of Prakṛti as ourselves and our lives circle around it. So long as we are enmeshed in mundane life, or even in its memories, experience of the underlying Puruṣa is not possible. Patañjali provides an eightfold method to overcome the hindrances: abstention *(yama),* observance *(niyama),* posture *(āsana),* breath control *(prāṇayama),* obliteration of senses *(pratyāhāra),* focused attention *(dhyāna),* contemplation *(dhārana),* and enlightenment *(samādhi).* The Buddhist path to enlightenment requires disentanglement from *saṁsāra,* or the phenomenal world. The Dhammapada describes the enlightened one as "free from the meshes of desires and the defilement of passions and free from all conditioning." (14.180)

Both the Buddhist and the Patañjali systems offer compatible recommendations. Abstinence from ill will or malice toward all forms of life are common to both. Jealously, greed, hatred, and anger are included among the undesirables, all of which have a narrowly defined self and ego as their basis. The gains and losses of past, present, and future will intrude into conscious awareness so long as they are important to the individual in question. Enlightenment has to become one's consuming passion, and it cannot coexist with desire for fame and fortune. As we saw in the previous chapter, abolition of "I" and "mine" are indispensable. The Buddha saw an egotistical self as the single serious impediment to *dharma.* Once self is erased, what remains will be *dharma,* the cosmic principle. In the Patañjali system, observances for internal and external purification, contentment, austerity, and devotion to God are recommended.

True selflessness consists of an expansion of self to include all other living creatures. This brings up the issue of meat eating and yogic practices. Older Upaniṣads certainly permitted and even recommended flesh eating (Bṛhadāranyaka Upaniṣad, VI, 4, 17–18). The Buddha also was not entirely clear on the point. The first of the five precepts of Buddhism enjoins the follower to abstain from killing. In Dhammapada, the Buddha says, "One is not noble who injures living creatures. They are noble who hurt no one" (19, 270). Yet it is unclear whether the Buddha firmly banned meat consumption. Elsewhere, he permitted meat eating provided the animal was not killed for that purpose (Vinayapiṭaka, Mahāvagga, VI, 31, 14).

Creation is not possible without destruction. Existence of one creature usually depends upon the destruction of another. Thus, it is impossible for a human to live without causing injury and death to other creatures. The real issue here is how sensitive one is to the issue and how hard one tries not to injure another living being. Instinctually, we do not eat creatures we love and with whom we identify. Cannibalism is considered despicable by most cultures. Most people would also refrain from eating monkeys because they resemble us, and most do not eat pet animals like cats and dogs.

It is obvious that loving and eating are basically incompatible. If we really and truly love all life around us, we will find eating all living creatures distinctly unpleasant. Fruits and vegetables and milk that nature creates for consumption may be the only exceptions. However, one may argue that even these contain life. There is no getting away from the fact that we have to destroy others' lives to sustain our own. The point here is to take life only when it is essential, while actively trying to minimize injury and pain. Killing for sport has no room in this mindset. Muslims take special care to slaughter animals in the kindest possible way, and slaughter is accompanied by prayer. The Native Americans ask permission of the deer before killing it for food.

In the absence of the desired changes in one's outlook on life, yoga will be shallow and very limited in scope. There will be little difference between yoga and common relaxation techniques and hypnosis. Yoga unsupported by ethical preparation and a yearning to know a reality beyond our senses is not yoga. The Buddha listed sensual attraction as one of the primary obstacles to enlightenment. The opposite of attraction, namely, aversion, was the second. One has to rise above the opposing currents of like and dislike, love and hate, and attraction and aversion. Restlessness and worry were pointed out as another major obstacle. Doubt and insufficient motivation will also get in the way. Although meditation depends upon quieting the mind, it should not be confused with laziness and torpor.

The Buddha, when he started out on his quest, initially took to meditation in isolation. However, after his enlightenment, he strongly supported living in a community of fellow spirituality seekers. He felt that the group would insulate the spiritual aspirant from fleshly temptations and reinforce resolve and determination. In addition, the warmth of mutual love and respect were conducive to spiritual progress. Opportunities to share spiritual insights through productive discussions, carried out in a spirit of friendship, were of great benefit.

It is easier to be selfless when surrounded by people we identify with, trust, and love. Maintaining moral and ethical values, character improvement, and community living are by no means unique to Patañjali and Buddha. Christians practice it in monasteries and convents, as do Sufis. Once Ānanda, his cousin, devoted attendant, and beloved pupil, said to the Buddha, "It would seem one half of the spiritual life is friendship with good people, association with good people, and communion with good people." The Buddha said in reply, "It is not so, Ānanda, it is not so. It is not one half of the spiritual life, it is the entire spiritual life" (Samyutta Nikāya, verse 2). The spirit of benevolence and love are indispensable even when meditation is practiced in solitude.

Here are a few words about physical preparation. The Buddha during the first years of his seeking enlightenment almost starved himself to death. In his first sermon at Sarnath he said:

> *Let me teach you the middle path, which avoids both extremes. By suffering, the emaciated aspirant produces confusing and sickly thoughts in his mind. Mortification is not conducive, even to worldly knowledge; how much less to a triumph over the senses! The sensual man is a slave to his passions, and pleasure seeking is degrading and vulgar. But to satisfy the necessities of life is not evil. To keep the body in good health is a duty, for otherwise we shall not be able to trim the lamp of wisdom, and keep our mind strong and clear.*

A balanced approach to physical and mental health is central to Buddhism. The Buddha regarded overeating, excessive sleep, and laziness as serious impediments to release (The Dhammapada, 23, 325). Patañjali took a similar view. He also did not think much of excessive sleeping and laziness. This sentiment was reflected in Gītā, which was written much later: "Those who sleep too much or too little, eat too much or too little cannot be a yogi" (VI, 16).

Tethering the Mind

Exercises designed to prepare the aspirant physically are an important part of the Patañjali system; in fact, the yogic exercises he emphasized are what his system is best known for. Patañjali recognized two types of yogic exercises: poses (the well-known lotus position, or *padmāsana*, for example) and breathing exercises. However, most of the currently popular yogic exercises did not come from Patañjali but from the yoga Upaniṣads composed later, sometime between 100 B.C. and 300 A.D. Yogatattva and Yogaśīva Upaniṣads, the best known of this group, subclassify different types of yogas and provide considerable details concerning different postures, breath controls, and the like. These Upaniṣads include Hatha yoga, which deals mainly with breathing exercises. The two Sanskrit letters, Ha and Tha, represent breathing through the right and the left nostrils, respectively. Rest and movements involve interactions between opposing groups of muscles. Poses, or *āsanas*, consist of a stationary and a dynamic phase of muscle action, followed by counterposes in which the opposing muscle groups are activated. These postures vary significantly in their complexity and purpose.

A variety of physiological studies are available on different poses, including standing on the head *(śīrṣāsana)*. Although claims have been made of yogis gaining volitional control over previously involuntary, autonomic functions, I am not aware of any solid scientific evidence that would support these claims. There is no doubt that yogic exercises produce physiological changes, but most of the changes seem to be brought about indirectly through the use of mechanisms under voluntary control. For example, there are several anecdotal reports of yogis being able to stop their hearts volitionally. However, the yogis were found to be using the respiratory muscles to bring about this effect by increasing the pressure inside the chest to minimize venous return to the heart. This, in turn, led to a significant drop in the pulse volume and heart sounds. The electrocardiogram continued to show heart contractions, although the pulse was feeble and difficult to detect.

Using various postures and breathing exercises, yogis are able to decrease energy requirements of the body to a bare minimum, and the body and the brain are brought into a stage of extreme rest and relaxation verging on inactivity. Experiments have been done on yogis being kept in airtight compartments for long periods of time. They were found to remain relaxed, without falling asleep, for ten hours or more. Drastic reduction in the oxygen content of the box to around 14 percent did not cause any discomfort. Outlandish claims of electrocardiographic silence during yogic trances and volitional control of body temperature, especially toward heat, and floating up in the air in defiance of gravitation (levitation), have also been made. The supportive evidence on these claims is highly questionable.

Breathing exercises are part of both the Patañjali and the Buddhist systems. However, Patañjali does not consider it essential. He refers to it as an optional measure but acknowledges its steadying influence on the mind. In other systems of yoga, such as the Hatha yoga system, breath control is central. Breathing is often used as a concentration point to prevent the mind from wandering. Blood and brain levels of life-giving oxygen and carbon dioxide are very sensitive to breathing, and profound changes in these gases can produce alterations in brain function and consciousness. Mountaineers, for example, report euphoria at very high altitudes, where oxygen content of the air is low. Some esoteric yogic exercises produce altered consciousness by manipulating blood gases through marked changes in respiration.

The main goal in all forms of meditation is silencing the mind. First and foremost, sensory input has to be minimized. A quiet, dimly lit environment is definitely beneficial, but perpetual stimulation cannot be avoided completely. One simply has to evolve a mechanism to disregard signals from the environment. In most systems, including the Patañjali system, withdrawing attention from the environment and the body to an internal locus in the mind is recommended. Since the human mind can process only one signal at a time, total attention to one will automatically mean withdrawal from all others. While some use chants or mantras to focus their attention, more experienced practitioners can direct it inward with no clearly defined focus. Under such conditions, the yogi becomes oblivious of, and detached from, all sensory inputs. Vālmīki, the author of Rāmāyaṇa, is believed to have hosted a termite mound around him during his protracted period of meditative trance.

Buddhists use a technique called mindfulness to deal with the report of the senses. This technique is by no means restricted to senses; it can be extended to any mental activity, including feelings, thoughts, and so forth. These phenomena are noted, but they are not cognized. The aspirant notes the experiences without accepting or rejecting them, totally passively, almost like an automaton.

After we manage to block out sensory signals, we have to deal with our thoughts and emotions. The restless, roving mind has to be tamed and calmed and brought into a single focus. This is the most difficult part; it calls for a great deal of patience and perseverance. A detached lifestyle definitely helps. Deemphasizing the ego and its needs and wants contributes substantially. Reduction in, and freedom from, sensory stimuli that send the mind drifting into the fantasy world are essential. The mind is normally itinerant; it is not at all easy to tether it to a single spot. Total absorption into a single internal focus helps the aspirant accomplish this.

As was stated previously, in the beginning the aspirant may use mantras. Others concentrate on the movements of the air through the nostrils. Success in eliminating distractions is usually heralded by luminous zigzags, spots, and sparks that pierce the darkness that sets in when eyes are closed. Ringing, buzzing, and chiming may be heard, and later, overt hallucinations may burst through. The beginner may find such experiences frightening.

While some Western authors regard such occurrences as ominous portents of a possible psychotic break, most experienced practitioners, both Eastern and Western, regard them as milestones along the path away from the perceptual world. These experiences are in many ways similar to those associated with sensory deprivation. As we leave the mundane world behind, we lose contact with all familiar points of reference. There is often a feeling of floating in an ocean of timelessness. The experience can be frightening to the beginner. Language is left behind, and we are immersed in recurring waves of ineffable emotions, mood states, and occurrences. Before we consider the train of events and experiences culminating in more elaborate, beatific, and illuminating states, we need to verify and establish the basic premise, namely, total inactivation of the mind.

Release from animal passions and longings of the ego calms the raging tempests of the mind and allows the aspirant to rest and relax. When such a person attempts meditation in a quiet environment in a natural setting, it will obviously lead to profound relaxation. Relaxation is an important step in the meditator's progress. This has been verified by a number of investigators, with various physiological techniques. A number of meditative techniques, including Zen, yoga, and transcendental meditation, have been investigated. These physiological changes suggested decreased activity of the sympathetic nervous system, which is responsible for the well-known fright-flight response. Significant decreases were observed in oxygen consumption and carbon dioxide elimination. Heart rate, respiratory rate, and arterial lactate concentration dropped. There was stabilization of muscle flow and decreased palmar perspiration.

Most investigators could not find significant changes of systolic and diastolic blood pressure and temperature. Several investigators documented decreased muscle activity as measured with electromyogram. While these findings are of interest, meditation is not primarily meant to produce relaxation. The relaxation response is simply a by-product. There are any number of techniques unrelated to spirituality, including hypnosis, progressive relaxation, autogenic training, and biofeedback, which produce comparable relaxation. Meditation, unlike the others, is designed and developed to take us away from the mundane world of time and space into another arena. While it is difficult, if not impossible, to prove the experience of the new reality, detachment from the mundane one is easier to establish.

Scientific Studies of Meditation

There are a number of problems with most available scientific literature on this topic. Relatively few studies have been done on Zen and Rāja yoga; most studies are on transcendental meditation. While most Zen and yoga practitioners in Japan and India, respectively, use the technique for their original purpose, namely, *nirvāṇa* and *samādhi*, Western practitioners use these techniques for a variety of purposes. As was mentioned previously, these techniques are ideally used as an adjunct to substantial changes in lifestyle, occupation, and worldview.

Many studies, especially on transcendental meditation, were conducted in people who used meditation without instituting the above-mentioned general, but important, lifestyle changes. Yardsticks of progress in meditation are exclusively subjective; there are no objective scales with which it can be verified. This is especially a problem with transcendental meditation studies where participants varied significantly in their interest, dedication, and duration of involvement.

A few recent, small-scale studies have looked at the effects of meditation on brain metabolism, but the sample sizes were too small to draw any firm conclusions. The electroencephalogram was the most common and most productive investigative technique used. Changes suggestive of deep relaxation involving the alpha rhythm (8–13 Hz), were the most common finding. This wave band is usually found in the normal waking state, with the subject sitting relaxed with eyes closed. During meditation, higher alpha densities were recorded. In addition, J. P. Banquet from the Stanley Cobb Laboratories for Psychiatric Research reported synchronization of the alpha activity across different parts of the brain. The meditators were able to maintain alpha activity (suggestive of uninterrupted reverie), even at the end of meditation, even with eyes open.

Usually EEG waveforms slower than alpha, called theta frequencies (6 to 7 Hz), suggest drowsiness and sleep. Subjects can be brought out of drowsiness and theta by using click sounds. Banquet found theta during meditation, in some subjects. Click, applied during meditation, abolished theta only for a few seconds; theta trains returned. This suggested meditation was associated with even less sensitivity to sensory stimulation than during sleep. EEG findings indicative of deep relaxation and detachment during meditation are not surprising. However, Banquet did come up with some surprises. He found trains of fast electrical activity of the brain suggestive of increased activity, breaking through the calm equipoise of meditation.

With deeper levels of meditation beta, waves at around 20 Hz (suggestive of activation) were present all over the scalp. Banquet asked his subjects to signal major changes in their subjective experiences with a button. Deep meditation and transcendence, as reported by the subjects, correlated with the appearance of beta rhythm. N. N. Das and H. Gastaut also reported similar high frequency patterns in yogis during deep meditation.

It is conceivable that under normal conditions, the superficial layers of the mind *(kośas)* suppress the underlying consciousness. Deep meditation might release consciousness from this inhibitory influence by suppressing the superficial layers.

The studies by B. K. Anand and associates from the All India Institute of Medical Sciences in New Delhi and A. Kasamatsu from the Department of Neuropsychiatry, Tokyo University Branch Hospital, are of special interest because the participants were experienced Indian and Japanese practitioners. Kasamatsu and associates studied EEG in three groups of Zen practitioners. Group 1 consisted of disciples with one to five years of experience, Group 2 of disciples with five to twenty years of experience, and Group 3 of priests with over twenty years of experience. They used research fellows and elderly men with no experience in Zen meditation as controls.

Like the other investigators, Kasamatsu found alpha waves during the initial stages of meditation. Within seconds of meditation commencement, well-organized alpha waves appeared in all brain regions and continued for several minutes, even with eyes open. Finally, rhythmical theta train made its appearance. Theta waves appeared only in some cases.

During the investigation, some disciples became drowsy, which allowed comparison of EEG during drowsiness and meditation. As we have seen, click stimulus alters the drowsy EEG pattern into alpha of arousal (wakefulness). Like Banquet, Kasamatsu also found that theta associated with meditation was disrupted by the click stimulus for a few moments only. Normally, when the stimulus is repeated, subjects habituate to its alpha-blocking effect. In other words, as the sensory stimulus is repeated, the subjects get used to it, and it will have less and less of an effect on the EEG. The alpha-blocking time is longer during first presentation of the stimulus, and it decreases in duration and almost disappears after the third or fourth stimulation. In Zen meditation, alpha blocking was less susceptible to habituation. The Zen masters did not get used to it, and they reported even clearer perception of the sensory stimulus during meditation than in their ordinary wakefulness.

Normally, as sleep sets in, alpha waves recede, and spindle-shaped EEG waves called sleep spindles and slower waves in the delta range appear. These electroencephalographic changes did not occur in meditation, and consciousness was not dulled as in sleep.

Anand studied four yogis who practiced Rāja yoga. EEG results were recorded before, as well as during, meditation. In this study, they used stimuli that were stronger and even noxious compared to those used by Kasamatsu. The subjects were exposed to strong light a loud banging noise, contact with a hot glass tube and ice-cold water, and a tuning fork. The effects on their EEG data were studied both before and during meditation.

Even under resting conditions, all yogis showed prominent alpha activity. During deep meditation *(samādhi)*, persistent alpha activity of high amplitude was seen (Figure 9.1). During meditation, external stimuli—namely, the tuning fork, strong light, and banging sounds—had no blocking effect of the alpha rhythm (Figure 9.2). Immersing the hand in cold water and contact with a hot glass tube were not perceived, and there were minimal changes in EEG (Figure 9.3).

Together, these studies allow certain conclusions. Meditative states of mind are different from sleep. Such EEG changes as sleep spindles, which characterize transition from wakefulness to sleep, are not seen during meditation. Theta persisting during eyes open is definitely not seen in subjects after they wake up from deep sleep. Beta waves found during deep meditation are never seen during deep sleep. Click stimulation during sleep blocks theta, which

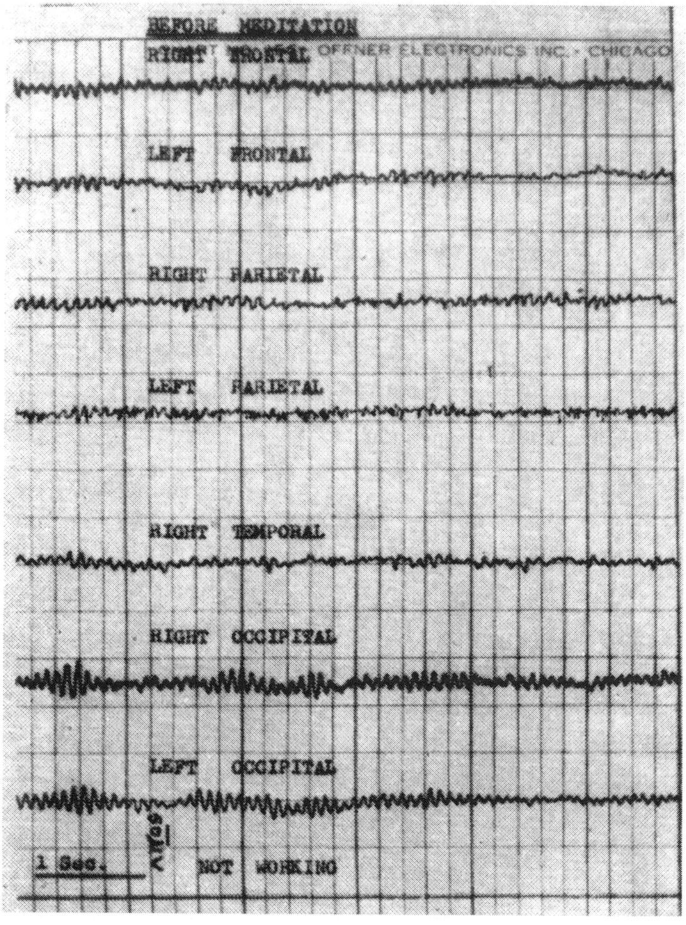

FIGURES 9.1 A AND B. ***Monopolar EEG scalp recordings of Śri Ramanand Yogi before meditation and during meditation.*** *Reference electrodes on both ear lobes were joined together. During meditation there was a well marked increased amplitude modulation of alpha activity, especially in*

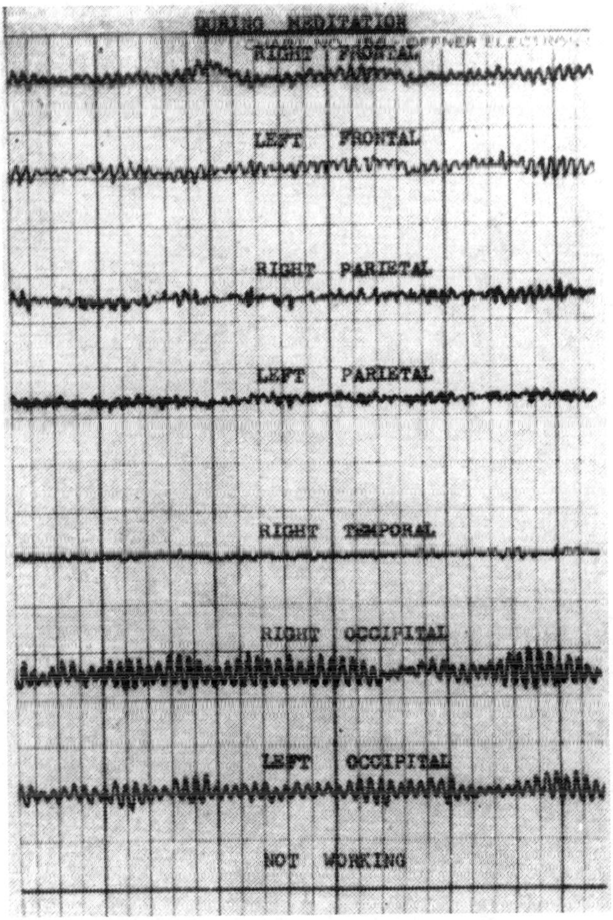

the occipital leads, where the amplitude increased from average to a maximum of 50–100 μV. The frequency was between 11–1/2 c/sec and 12 c/sec both before and during meditation.

Figure 9.2A

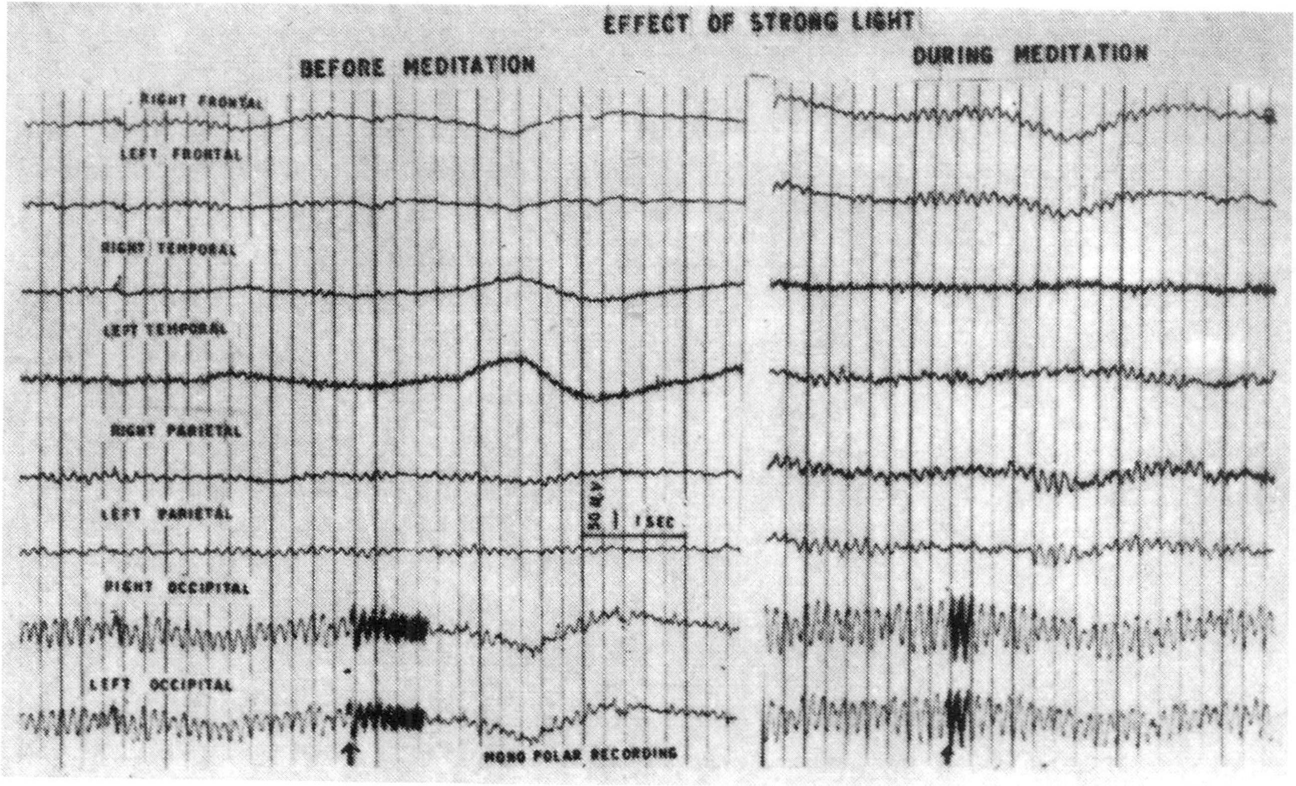
EFFECT OF STRONG LIGHT
BEFORE MEDITATION
DURING MEDITATION
RIGHT FRONTAL
LEFT FRONTAL
RIGHT TEMPORAL
LEFT TEMPORAL
RIGHT PARIETAL
LEFT PARIETAL
RIGHT OCCIPITAL
LEFT OCCIPITAL
1 SEC
MONO POLAR RECORDING

FIGURE 9.2B

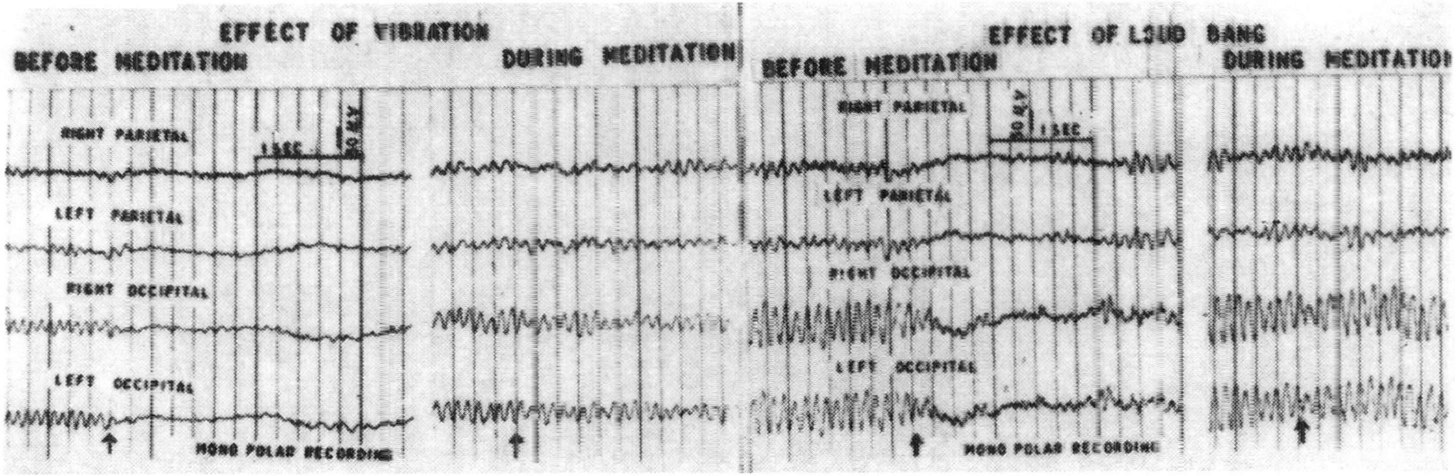
EFFECT OF VIBRATION
BEFORE MEDITATION
DURING MEDITATION
RIGHT PARIETAL
1 SEC
LEFT PARIETAL
RIGHT OCCIPITAL
LEFT OCCIPITAL
MONO POLAR RECORDING
EFFECT OF LOUD BANG
BEFORE MEDITATION
DURING MEDITATION
RIGHT PARIETAL
1 SEC
LEFT PARIETAL
RIGHT OCCIPITAL
LEFT OCCIPITAL
MONO POLAR RECORDING

FIGURE 9.2C

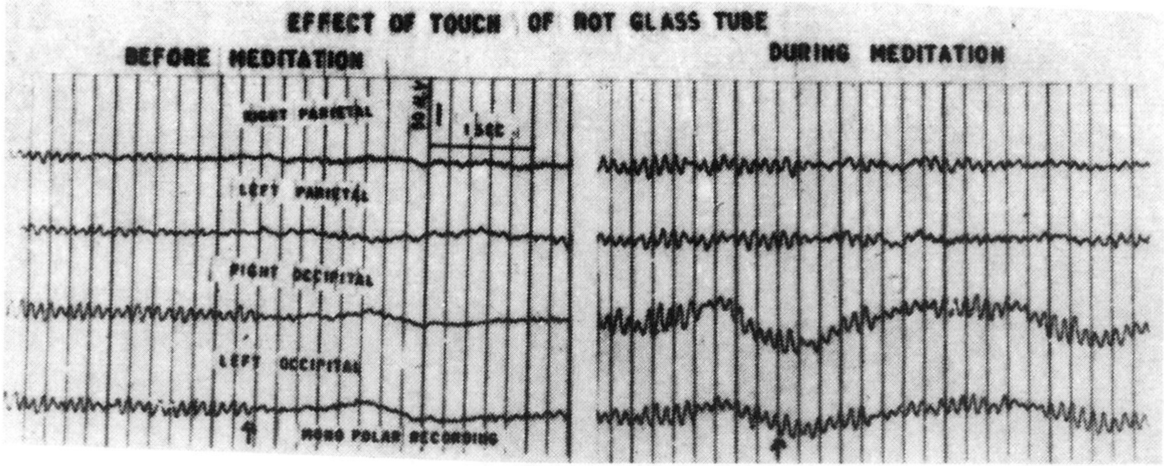

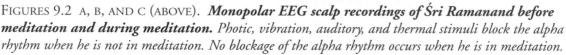

FIGURES 9.2 A, B, AND C (ABOVE). ***Monopolar EEG scalp recordings of Śri Ramanand before meditation and during meditation.*** *Photic, vibration, auditory, and thermal stimuli block the alpha rhythm when he is not in meditation. No blockage of the alpha rhythm occurs when he is in meditation.*

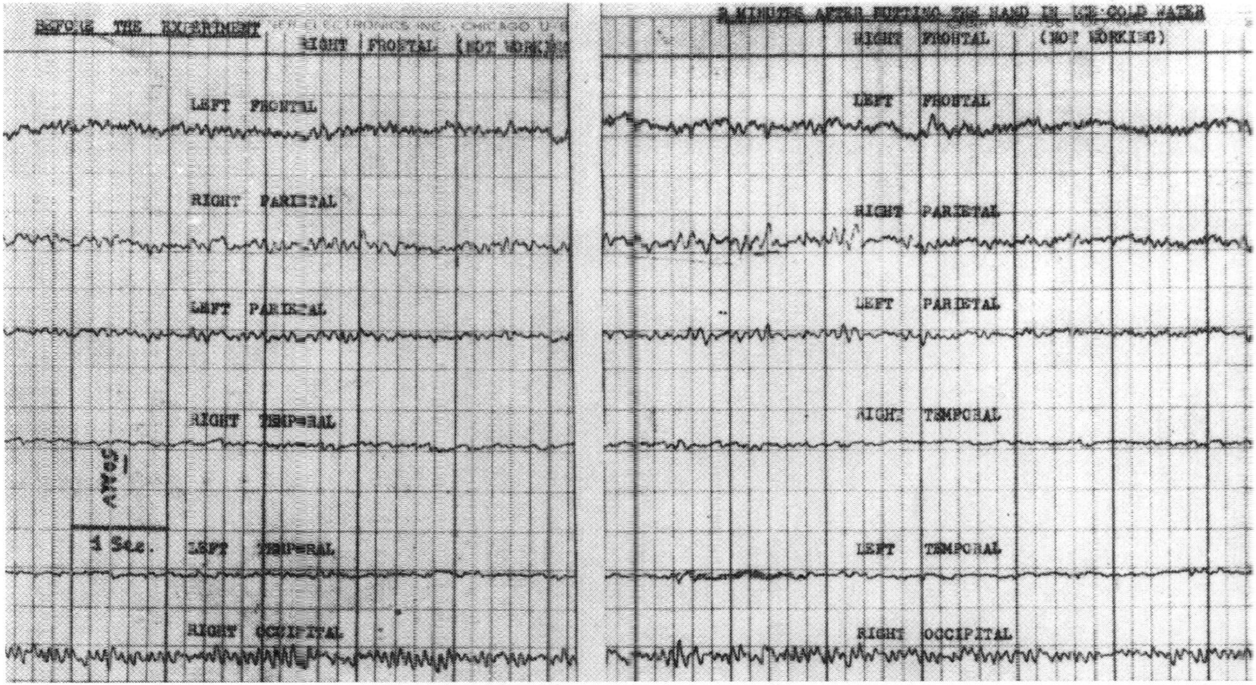

FIGURE 9.3 (BELOW) ***Monopolar EEG scalp recordings of Śri Babu Ram Gupta who kept his right hand immersed in cold water at 4°C for 55 min.*** *It shows records taken before the start of the experiment and 3 minutes, after placing the hand in water. The alpha activity of 11 c/sec persisted throughout the whole period the hand was kept in water. No discomfort or pain was felt by him throughout this period.*

is replaced by alpha; click stimulus during meditation blocked the EEG pattern only momentarily with the return of theta. Lastly, the participants reported totally different subjective experiences during meditation, compared to sleep.

Meditation is designed to silence the mind so that the uncontaminated consciousness becomes the summum bonum of the mental experience. These studies provide objective proof for this. *Kośas,* or sheaths, discussed in Chapter 4, cloak consciousness. Lack of subjective or EEG responses to sensory stimulation indicates that the perceptual apparatus becomes silenced. The location of inactivation in the sensory data processing mechanism is different for Rāja yoga and Zen. In the study of Rāja yoga, sensory stimulation elicited minimal if any EEG response. This indicates that the brain did not acknowledge the sensory input at all. The Zen practitioners, on the other hand, showed response, but unlike normals, they failed to habituate, indicating inactivation of the sensory processing routine. In mindfulness of Zen meditation, sensory stimuli are dealt with in a manner diametrically opposite to that of Rāja yoga. The practitioners take note of everything that happens within and without, but do not process it. Thus, the mental mechanisms responsible for signal reception and processing are inactive in Rāja yoga and Zen, respectively.

The study by Anand showed no subjective or EEG response to a hand immersed in cold water kept at 4 degrees centigrade for more than thirty minutes and to touching with a hot glass rod. This clearly shows switching off of body perception. Vital functions such as heart rate and respiration show significant slowing. The final goal in both systems is disabling the empirical mind. Alpha waves of increased amplitude during meditation while the subjects were awake indicate profound states of relaxation. Deep meditation was associated with even slower brain waves in the theta range. Yet the subjects were awake. An EEG pattern comparable to that of sleep when the subject is wide awake indicates a state of mind isolated from all activities of the mind. Thus, the scientific data available indicate a state of mind in which consciousness is the sole activity (Figure 9.4).

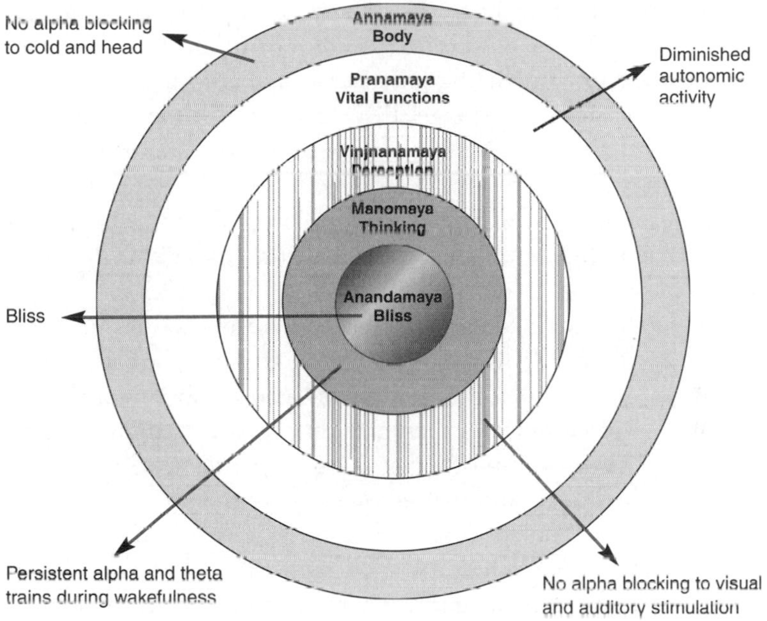

FIGURE 9.4 *Different layers of the mind and scientific evidence in support of their inactivation during meditation.*

Transcendence

According to Kaṭha Upaniṣad, "Beyond the senses are the objects (of the senses) and beyond the objects is the mind; beyond the mind is the understanding and beyond the understanding is the great Self" (I, 3, 10). "When the five senses together with the mind ceases and the intellect itself does not stir, that, they say, is the highest state" (Kaṭha Upaniṣad, II, 3, 10).

Most investigators did not, or could not, evaluate the subjective aspects of such a state of detachment and isolation. S. Venkatesh from the National Institute of Mental Health and Neurosciences in Bangalore, India, studied the phenomenology of consciousness in twelve senior Kuṇḍalinī meditators. Kuṇḍalinī meditation is based on an esoteric anatomy of the energy systems of the body. According to this system, there are six or seven energy centers along the spinal cord. At the lowermost center, Kuṇḍalinī, the serpent power, lies dormant. By the practice of Hatha yoga, Kuṇḍalinī is awakened. It raises up through the central canal of the vertical axis, piercing the different centers en route to the uppermost center at the vertex. Union of the Kuṇḍalinī power with the last center represents the union of opposing forces (Yin and Yang) of the phenomenal world. This union is associated with an outpouring of divine ecstasy.

Venkatesh found altered experience in perception, meaning, and time sense during the meditative trance. The practitioners reported increased joy and love. The results indicated a substantial shift in the quality of conscious experience during the meditative state. Language, however, and the rating instruments based on it, can provide only limited information on such states.

Many authors, notably the Buddhists, have provided more elaborate descriptions of the subjective experience. Thought is normally conditioned by language. As meditation deepens, language and its products are totally abandoned. Initially, the mind is contaminated by periodic incursions of thoughts and emotions. This mental state is without qualities of any sort and simply cannot be reduced to language. Here, the mind comes to a total standstill, experiencing no undulations. Time and space barriers are totally transcended and the final experience is untainted by either. Hindus call this state *turīya,* the fourth state, and Buddhists, *śūnyatā*.

Although all traditions acknowledge the ineffability of this state, many efforts to describe it have been attempted. They provide some clues on what it might be like, but at best they are approximations. Śāṁkara's unrivaled ability to provide lucid descriptions of the indescribable and to provide penetrating explanations for the inexplicable earned him the title Ācārya (most revered teacher). I find some of Śāṁkara's comments about the state of mind associated with enlightenment particularly helpful and useful.

When the subject becomes the sole object that occupies the mind, it ceases to be either subject or object. The subject-object distinctions completely disappear. Thus, the knowing subject, the object of knowledge, and the process of knowing all become fused into a single whole. In an objectless experience with no points of reference, time and space are irrelevant. Change is alteration over time, and where there is no time, there are no changes, no contrasts, and no predications. This would seem to correspond to the Buddhist concept of emptiness, or *śūnyatā*.

Although the experience itself cannot be described, it is possible to comment on the individual's response to the experience. However, even this will have to be an approximation. First and foremost, the individual is convinced beyond doubt that this experience is the ultimate reality, unlike anything experienced before. It has been compared to the realization of the unreality of dreams upon waking up. The realization is immediate and conclusive. Vedas refer to it as *sat,* for which there is no simple equivalent in English. *Sat* stands for that which is essential, fundamental, and most substantial. It cannot be reduced, divided, or contrasted. The closest English expression is absolute truth, which is different from relative truth or honesty. It is the Truth Jesus spoke of when he said, "I am the way, truth and life" (John, 14.6). According to St. Augustine, "Where I found Truth, there I found my God, the Truth itself." According to the Qur'an, "God is Truth, whatever else they call upon is false." Gandhi used to say, "Truth is God, and not God is truth."

The experience is usually associated with a sense of illumination. The illumination can be luminance in a physical sense, or a brightening of the mind. Paul, during his conversion experience, talked about "a light from heaven." According to John, "While you have the light, believe in the light, that you may be the children of light" (12.35). According to Chāndogya Upaniṣad: "Now that serene being, out of body reaches the highest light. This is his real self. This is the immortal, the fearless Brahman, whose name is Truth" (VIII, 3, 4). A light is supposed to have appeared above the Buddha's head at the time of his enlightenment. In Dhammapada, the Buddha said, "They leave darkness behind and follow the light" (6, 87). One of Prophet Muhammad's most famous prayers is known as the Prayer of Light: "O Allah, place light in my heart, light in my sight, light in my hearing, light on my right hand and on my left, light above me, light below me, light behind me and light before me. O Allah, Who knows the innermost secrets of our hearts, lead me out of darkness into Light."

More recently many individuals, including Bill Wilson, the founder of Alcoholics Anonymous, have described a sense of illumination during spiritual experiences. Einstein's absolute also happens to be light.

The mood associated with the experience, at least initially, is bliss or joy. "Joy" should not be confused with mundane pleasure that contrasts with pain. Joy is unidimensional with no antipode. Qualitatively, it resembles the mood state associated with dreamless sleep. As meditation deepens, spiritual ecstasy looms and engulfs the entire mind. The last meditative state is totally beyond the Yin-Yang duality and "joy" in a realm of uniformity, consistency, and purity. Buddhist literature makes many references to this extraordinary state of mind. The Dhammapada states, "Meditating earnestly and striving for nirvāṇa, they attain the highest joy and freedom" (2, 23). And again: "When a sage stills his mind, he enters an empty house; his heart is full of the divine joy of the Dhamma. Understanding the rise and fall of the elements that make up the body, he attains the joy of immortality" (25, 373–374).

As we have seen, this type of pleasure, unrelated to perception, is not triggered by external events. In his discussion with King Ajātasattu on the fruits of the life of a recluse, the Buddha explained and exemplified this phenomenon.

> *His very body does he so pervade, drench, permeate, and suffuse with the joy and ease born of concentration, that there is no spot in his whole frame not suffused therewith. Just, O king, as if there were a deep pool, with water welling up into it from a spring beneath, and with no inlet from the east or west, from the north or south, and the god should not from time to time send down showers of rain upon it. Still the current of cool waters rising up from that spring would pervade, fill, permeate, and suffuse the pool with cool waters, and there will be no part*

> *or portion of the pool unsuffused therewith.* (Sāmaṇṇphala Sutta, II, 77, 78)

Whether this "reward" is mediated by the norepinephrine that mediates arousal or by something beyond neurochemicals, we do not know.

Time and space undergird the world we live in. Our concept of reality is dependent upon time and space. In fact, in Newton's time, scientists subscribed to notions of absolute time and space. Although Einstein debunked these concepts and modern physicists have totally discarded them, most of us still believe in them. Anybody who believes in transcending time and space barriers is automatically regarded as eccentric; and that is putting it mildly. A time machine is more acceptable than a person who can transgress these boundaries. Admittedly, there is no objective evidence in support of such claims.

There are two main reasons for this. First of all, individuals who reject the phenomenal world have little interest in proving or disproving. True transcendence is a rarity. Of the many to embark, only a few reach such exalted levels. Most individuals with that degree of selflessness are indifferent to their own personal accomplishments and are silent about their experiences. Statistics can verify group phenomena but not individual ones. All medical journals of repute recognize this, and they publish anecdotal reports of rare diseases and unusual reactions and adverse responses to treatments. Such reports are accepted by scientific medicine. Transcendental experiences are not easy to come by, and anything more than anecdotes will be next to impossible.

Physicists tell us that notions about a plane beyond time and space are not all that far-fetched. According to the physicist Michael Kaku, "Many of the world's leading physicists now believe that dimensions beyond the usual four of time and space might exist."[1] In fact, a number of prestigious physics departments are active centers of research on higher dimensional space-time. We do not know for sure if human beings can acquire access to these higher dimensions; both the Buddha and Patañjali, and many others, answer in the affirmative. But they all mention an even higher plane beyond all dimensions, a plane where planes do not exist.

Buddhism differentiates between a momentary flash of enlightenment *(bodhi)* and true liberation *(nirvāṇa)*. The former may be accomplished through arduous meditation, or one may accidentally bump into it. The latter is possible only if the enlightenment experience is supported both by intellectual knowledge and desired personality changes. Enlightenment comes easier to some than to others. People who are soft-mannered, kind-hearted, and not bound by dogma and dictum are more likely to have the experience than narrow-minded fanatics.

Someone with a psychological bent that favors numerical reckoning, a linear style of information handling, and extreme reliance on logic might have the greatest difficulty. Artists, musicians, and nature-lovers who use a holistic style seem to find this easier. In Zen Buddhism, the enlightenment seeker is presented with insoluble intellectual problems called *kaons* to exhaust and extinguish the intellectual, linguistic style of thinking. A *kaon* is an oxymoronic puzzle designed to exhaust language-dependent mental faculties.

According to Indian thought, consciousness is synonymous with the uncreated Absolute. It predates and exceeds science and objectivity. True subject, which is what consciousness ultimately is, can never be moved into the objective space and, therefore, can never be subject to objective research. Science is secondary to consciousness and, therefore, it cannot be used to prove or disprove it. Even if the brain were destroyed, the primordial element within should remain as such. Consciousness in this sense is independent of, and prior to, life. Verification of this would require contact with the dead and it would indeed be difficult.

However, there are a number of well-documented reports of near-death experiences. These are subjective reports provided by people who were pronounced dead but later came back to life. In all cases, physiological signs of life including EEG indicated brain death. Deaths were due to a variety of causes including accidents, childbirth, surgery, sudden illness, criminal attacks, etc. While a minority of these individuals had unpleasant experiences, the post-death experiences were pleasurable for most. A number of subjects gave explicit descriptions of the events that followed their deaths. Other people who were present in the room corroborated their descriptions of what happened in the immediate environment after they were pronounced dead. In several cases, they were able to describe parts of the hospitals that they had never previously visited while they were alive.

In many ways, there are striking similarities between these reports and descriptions of deep meditative trances. First of all, several subjects commented on the ineffability of their experiences. As mentioned previously, a sense of illumination is recognized both by the Upaniṣads and Buddhists. Dr. Raymond Moody states in *Life After Life:* "What is perhaps the most incredible common element in the accounts I have studied, and is certainly the element which has the most profound effect upon the individual, is the encounter with a very bright light . . . Even though this light is of indescribable brilliance, many make the specific point that it does not in any way hurt their eyes or dazzle them."[2]

Several described a sense of beauty. Many reported a feeling of "reality" with a sense of clarity that was not there when they were alive. *Sat,* or truth, is a fundamental characteristic of reality. The only way to know it is to experience it. Many reported a feeling of bliss, a sense of happiness they had never known before. Once again *ānanda,* according to all religions, is part of divine experience.

Near-death experiences of panoramic visions of the past are reminiscent of the Buddha's descriptions of his past lives and those of others that dawned upon him during his enlightenment. It is difficult to differentiate between remembering scenes from the past and transcending the time barrier to shift into the past. The following anecdote from *Return from Death* by Grey combines near-death experience with several factors associated with deep meditative trance: light, beauty, peace, and loss of identity. "I found myself in this extremely bright light and felt absolute peace. I felt the light and peace were one . . . I was suddenly in the light and it was beautiful. I had no sense of separate identity. I was in the light and one with it."[3]

We have discussed different ways of escaping the trammels of the flesh and moving toward reaching a transcendent, nonmaterial reality. There are many yogas we did not consider here. Which path to choose is a major dilemma facing the spiritual aspirant. What about those who are not interested in spirituality at all? Is spirituality indispensable and essential? Are we doomed if we have no interest in yogas at all? Two years ago when I was visiting the city of Madras in India, I happened to meet a retired judge, Mr. Krishnaswamy Reddy, who was large-hearted, broad-minded, and deeply erudite. To me, people like him represent Hinduism at its best. When I posed these questions to him, he cited a passage from the Bhagavad Gītā (18, 63) in reply:

I have revealed to you truths hitherto enshrouded in māyā
Study them carefully and in detail and then act as you see fit.

The Beginning

Mary Smith, a hopeless cocaine addict, was in the throes of total desperation and deep depression. Suicidal thoughts darkened her mind. Out of the blue, she had a profound experience: unexpected, unannounced, and inexplicable. The experience changed her life forever. She stopped abusing cocaine and became the responsible employee and the affectionate mother that she once had been. She was certain that she had had contact with the Divine.

Since the dawn of history, all over the world, large numbers of people have had similar experiences. Many were transformed by the experience and gained insights hitherto unknown.

The experience often transcended the mundane and gave the individual insights of great significance, not only for that person but also for the society as a whole. The origins of many leading religions of the world can be traced to these experiences.

Whether such experiences are "real" will depend upon our point of reference. Most regard the waking experience as real. However, philosophers and scientists alike question this viewpoint. For over 2,000 years, Indian philosophers have believed that the waking experience that rests on the slippery slopes of time and space is not substantial. The phenomenal world displayed during waking hours is not nonexistent, like the son of a barren woman or a flower in the sky. It is the true reality, but time and space disguise it. Since Einstein, scientists, too, have recognized time and space as relative. We realize the unreality of dreams—also bound in time and space—when we wake up. Similarly, we have to wake up from our normal state of consciousness to experience the true reality within. Mary Smith had such a rare moment of true wakefulness.

Mary was convinced that she was in the presence of the Lord. *Sat,* or absolute truth—not to be confused with honesty, which is relative truth—is a fundamental characteristic of such experiences. This is reality uncontaminated by time, space, or matter. A sense of illumination is common. The Upaniṣads call the Absolute "*jyotiṣāṁ jyotiḥ*"—light of lights (Bṛhadāranyaka Upaniṣad, IV, 4, 16). By coincidence, Einstein's absolute was the speed of light. Spiritual ecstasy or bliss is the crowning glory of the experience. This sense of exquisite happiness has no parallel in mundane life. It is totally unlike the pleasures associated with acquisitions and accomplishments. The entire experience is ineffable.

Since the brain is the seat of the mind, it would make sense to look for relevant brain mechanisms. An experience that was clearly subjective should have an objective basis in the brain. As a matter of fact, both the internal subjective and external objective world should have the brain as the basis. The external is created by one region of the brain, while another produces the internal; both are its productions.

Diseases of the brain, notably epilepsy, can produce unusual states of mind. In fact, several treatises are available on the relationship between temporal lobe epilepsy and spiritual states of mind. Attempts have been made to explain the theophanic experiences of the Buddha, St. Paul, Muhammad, Joseph Smith, Joan of Arc, Emanuel Swedenborg, and others as seizure manifestations. These reports vary widely in quality. In one report under the unflattering title of "G. Buddha"—which is similar to "J. Christ"—the author gave as the reason for his conclusion that "it is a fact that the Buddha suffered from seizure disorder since childhood": His mother carried the pregnancy for a full ten months; when he was an adolescent, once when no one was watching, the Buddha sat down cross-legged and went into a trance; and he had altered consciousness during his enlightenment. I find the justification unconvincing. Some of the other authors are quite thorough, and they have painstakingly searched ancient literature to salvage supportive findings. However, there are still a number of problems with their claim.

True epileptic seizures with religious overtones are extremely rare. Epilepsy, associated with euphoria, is even more rare. It would seem most unlikely that the saints and sages referred to above and a sizable percentage of the people who report similar transcendent experiences—meaning 40 to 60 percent of the people in the United States, Great Britain, and Australia—have had such experiences. Epileptic seizures are often, if not always, associated with confusion and subsequent amnesia. Most saints and sages claimed extraordinary clarity of mind during the spells, and the experience, with all its details, was etched into their memories. None of the methods they recommend to reach such spiritual states even vaguely resemble methods for seizure induction. Many unsung followers of these saints and sages retraced their steps to get to the desired goal, and none of them had a seizure disorder. Whether some of them had epilepsy we do not know, but, epilepsy, as an explanation for their spiritual experiences and insights is seriously deficient.

Mary Smith definitely did not have a seizure disorder. Neither she nor anyone in her family suffered from epilepsy. During the episode, she did not have tremors, shakes, or clonic movements of any sort. She was not confused or disoriented either during or after the experience. She had no discomfort of any sort, including headaches; in fact, she was unusually happy.

The inadequacy of epilepsy as an explanation does not exclude other neurological possibilities as well. The mind is a mosaic of multifarious functions. Ancient Indians conceptualized it as layers representing the body, vital functions, conation, cognition, and so on that encircle a central core. The last layer intimately associated with the radiant center is *ānanda,* or bliss. The central core has both empirical and transcendental components. The former has a neurologic basis, but the latter does not. Transcendental consciousness is the Absolute, and realizing it brings enlightenment.

The transcendental core beyond time and space represents the Primary Principle. Time and space and everything that depend upon it have their basis in it. It is surrounded by created elements of the brain and the mind. The veils that encircle it are sediments of millions of years of evolution, and enlightenment is retracing the steps of evolution back into the uncreated Absolute at the center. The word "perception" has no application here. This is a process of confluence or union, of yoga.

There are a number of ways of accomplishing this. The creations carry the Creator's mark. With the right mindset, the Creator can therefore be accessed through the creations. Hinduism recommends four paths: spiritual knowledge, or Jnāna; devotion, or Bhakti; conduct, or karma; Rāja yoga, the fourth, provides direct access through contemplation. There exist any number of unnamed yogas. Numerous rivers, many without names, find their way to the ocean.

Reality as we know it has time and space as its basis. As we transcend time and space, phenomenal reality begins to soften, melt, and run. As we accelerate close to the speed of light, reality as we know it, and the world therein, will disintegrate and dissolve.

Both Patañjali, the most respected Hindu authority on meditation, and the Buddha, refer to the acquisition of supernatural abilities, when time and space are transcended. While some traditions use yogic techniques specifically to obtain these supernormal abilities, the Buddha, Patañjali, and other straight-liners look down upon it. They regard these abilities as distractions and remind us that people who get sidetracked by them never reach true liberation. People who seek fame and prosperity through these techniques are as helplessly bound to the mundane world as anybody else, and they have very little to teach.

Once a rich man in Rājagṛha, where the Buddha resided, had an exquisite bowl carved from the finest sandalwood, and he challenged the monks to get it from the top of a tall, inaccessible bamboo pole. Bhāradvaja, an accomplished *arhat* (enlightened monk), levitated into the air, got the bowl, and presented it to the Buddha. The Buddha rebuked him, saying, "Like a woman who displays herself for the sake of a miserable piece of money, have you, for the sake of a miserable wooden pot, displayed before the laity your power to make miracles?" He pointed out that such behavior would benefit neither the performer nor the audience. He declared miracle-making a punishable offense (Vinayapiṭaka, Cullavagga, V, 8, 2).

According to Vivekananda: "By means of concentration the mind may become such a fine instrument that its possessor may at times develop certain supernatural powers like that of seeing into the future, or reading another's thoughts. Although such powers in themselves have hardly any spiritual value, the person who has acquired them may easily succumb to the temptation of making a business out of them."[1]

Physicists tell us that notions about a plane beyond time and space are not all that far-fetched. We do not know for sure whether human beings can acquire access to these higher dimensions; both the Buddha and Patañjali, and many others, answer in the affirmative. But they all refer to an even higher plane beyond all dimensions, a plane where planes do not exist.

Enlightenment that comes with effort may also come spontaneously with no effort as happened to Mary. Similarly, paranormal abilities may come spontaneously and naturally to some, like my mother. She was not the only one who believed in prescient dreams. Both Śāṁkara and the Buddha acknowledged precognition associated with dreams.

Birth and death are phenomena bound to time and space. Only the superficial appurtenances of the mind are born and only they decay and die. It is normal and natural to fear death. However, the basis for such fear is our concept of existence as limited to matter and its creations. Glimpses of the spirit beyond matter will diminish the fear and the pain that accompanies dying and death.

Humans are basically hedonistic. All our actions are directly and indirectly pursuits of pleasure. Of the many pleasure chemicals of the brain, dopamine appears to be principal. The professional success, wealth, stable marriage, and warm family we strive for—these things give us pleasure. At first glance, all pleasure may seem the same. However, on closer scrutiny, the pleasure associated with making money would appear distinct in quality from that associated with enjoying a panoramic view from the mountaintop. While most people are able to appreciate the difference between the two, articulating the difference is not easy.

Life may indeed be pleasure-driven; however, there appears to be more than one type of pleasure. We know of at least two: one associated with personal success and gratification and the second associated with aesthetic rapture. This means that we have a choice in the type of pleasure we seek.

The two types of pleasures are mediated by different brain mechanisms. Language is predominantly a dominant hemispheric function. The dominant hemisphere also uses a linear modus operandi. The nondominant hemisphere, on the other hand, is holistic and uses nonlinguistic modes of communication. There are dopamine pathways that mediate pleasure in both hemispheres. Although the same neurochemical might be involved in mediating pleasures linked to both dominant and nondominant hemispheric activity, in quality they differ sharply.

Concept of self, conditioned by language and sharply separated from the surroundings, is likely to be a creation of the dominant hemisphere. "I" is temporally and spatially cleaved from "not I." Mundane pleasure and pain are more related to successes and failures, respectively, of a sharply defined "I." The holistic nondominant hemisphere is less able to separate self from surroundings.

Art, music, and nature are nondominant hemispheric pleasures, and they do not involve a well-defined concept of self. In fact, blurring of the boundaries between self and not self commonly accompanies the rapture of all aesthetic enjoyment. Being with family and friends, enjoying the idyllic beauty of nature and feeling bonded to our pets are all activities that involve dissolution of identity and bliss. In fact, the intensity of this type of pleasure runs in tandem with degrees of selflessness and absorption into a collective whole. Deep sleep associated with a total loss of identity is blissful. Total transcendence through intense meditation, people say, brings the ultimate happiness.

The well-formed concept of self is a product of evolution and its dissolution a product of devolution. The dominant hemisphere creates the well-demarcated self. Egoistic pleasure mediated by the dominant hemisphere has pain as its antithesis. The nondominant hemispheric self, less sharp and more fluid, produces a pleasure that does not have pain as its opposite. The ultimate pleasure Indians believe is *ānanda,* the experience of uncontaminated consciousness—God.

Pure consciousness is before and beyond the phenomenal world. The transcendental pleasure of consciousness, however, is not linked to things and accomplishments, it just is. It has no polar opposite. Here there is no self.

Pleasures unrelated to a narrow concept of self are closely associated with the term "spirituality." This includes pleasures that accompany art, altruism, devotion, and meditation. Spirituality can be found in all actions, provided the motive is selfless. Here the pleasure is a by-product; the primary goal is yoga, or union, with oneself, with nature, and with God. This calls for a certain mindset that the Upanisads call *Upāsana*. Literally, the word means "sitting near"; philosophically, it has a much larger, wider, and deeper connotation. According to Śāṁkara, it is a vehicle of establishing union with the Absolute as firm as the union one feels with the physical body in ordinary life. It is looking for an inner meaning and substance in all experiences of life, or "searching for the thread within the thread."

According to ancient Indians, the cosmos undergoes endless cycles of creation and absorption. Evolution is not a linear but a cyclic process. Man stands at the threshold of bestial and spiritual existence. Unlike the animals, man is able to intuit the transcendental Absolute and to access it. Evolution that cleaved self from surroundings also developed the ability for the self to retrace its steps and to experience the beginnings.

The human mind contains both the dark shadows of our past and the bright promise of tomorrow. Within the folds of the human mind, there exist the passion of the animal, the cruelty of the savage, and the selfishness of the egotist. The mind also contains the warmth of love, the softness of compassion, and the generosity of giving. Instinctually, we anticipate, hope for, and look forward to the days of universal brotherhood. We are ashamed of our animal past and the behaviors associated with it. Covering organs of elimination and reproduction, which mark our links to the animal, is a hallmark of civilization. The modern man is ashamed to eliminate and procreate in public.

The human psyche appears to have taken on characteristics difficult to explain by the dicta of evolution as we understand it. Self-sacrifice for the benefit of others is considered highly desirable. We have become more concerned about the underprivileged, with fair play and equality seen as hallmarks of progress. We have become more concerned about the environment than ever before and more sensitive toward how we treat animals. Organizations such as the Society for the Prevention of Cruelty to Animals have sprung up, and every effort is being made to prevent rare species from becoming extinct. Most countries have enacted laws that protect animals from misuse and abuse. These traits herald expansion of a narrowly defined concept of self to one that embraces the living and the nonliving that surround us. This is our cherished hope; this is the long-awaited dream; this is the Promised Land.

Human history is bloodstained. In the past we utilized our strength, knowledge, and intellect to enslave, dominate, and exploit other humans and nature. Finally, the dark clouds of egotism are lifting, and the light of truth is breaking through. We have, however, a long way to go. Narrow-mindedness, dogmatism, and bigotry are by no means gone. They are both around us and within us. Our very survival depends upon our ability to overcome these pernicious characteristics.

We did not engineer the big bang; we did not set evolution in motion; we did not shape the laws of nature. Something beyond created us and brought us to where we are, and it will carry us further, to wherever we are destined. Life is unlikely to end with humans, even if we burn in a nuclear holocaust. The relentless wheel of evolution will pick up from where we leave off and roll to its predestined goal. If the human mind continues to evolve, enlarge, and expand, so that we are able to recognize our kinship with the creations around us, so that we are able to grasp our oneness with the cosmos, and so that we merge in yoga with the Divine, the long cosmic cycle will disclose its cryptic secret, and the long saga of billions of years of evolution will display its profound significance.

Notes

Chapter 1

1. J. Kornfield, *A Path with Heart: A Guide Through the Perils and Promises of Spiritual Life* (New York: Bantam Books, 1993), p. 288.

2. J. Nehru, *The Discovery of India* (New Delhi: Jawaharlal Nehru Memorial Fund, 1981), chap. 4, "What Is Hinduism?"

3. Ibid.

4. Ibid.

5. S. Radhakrishnan, *Indian Philosophy*, vol. 1 (Delhi: Oxford University Press, 1989), "Introduction: General Characteristics of Indian Thought."

6. P. Davies, *The Mind of God: The Scientific Basis for a Rational World* (New York: Simon and Schuster, 1992).

7. S. Hawking, *A Brief History of Time: From the Big Bang to Black Holes* (New York: Bantam Books, 1988).

8. Ibid.

9. R. Leaky and R. Lewin, *The Sixth Extinction: Pattern of Life and the Future of Humankind* (New York: Doubleday, 1995).

10. Hawking, *A Brief History of Time*.

11. S. K. Chatterji, "Linguistic Survey of India: Languages and Scripts," in *The Cultural Heritage of India*, 2d ed., vol. 1: *The Early Phases*, ed. S. K. Chatterji, N. Dutt, A. D. Pusalker, and N. K. Bose (Calcutta: Ramakrishna Mission, 1958), p. 70.

12. C. Sagan, *Cosmos* (New York: Random House, 1980), p. 258.

Chapter 2

1. A. Parthasarathy, *Vedanta Treatise*, 4th ed. (Bombay: Vedanta Life Institute, 1992), pp. 17–18.

2. R.T.W. Davids, *Dialogues of the Buddha,* 1st Indian ed. (Delhi: Motilal Banarsidass, 2000), Sāmaṇṇphala Sutta, verse 55.

Chapter 3

1. S. Hawking, *A Brief History of Time: From the Big Bang to Black Holes* (New York: Bantam Books, 1988).
2. S. Radhakrishnan, *Indian Philosophy,* vol. 1 (Calcutta: Oxford University Press, 1923), p. 95.
3. Vivekananda, "The Absolute and Its Manifestations," lecture delivered in London, 1896.
4. M. Frydman, *I Am That: Talks with Sri Nisargadatta Maharaj,* 2d ed. (Durham, N.C.: Acron Press, 1976).

Chapter 6

1. R. E. Schultes and A. Hofmann, *Plants of the Gods: Their Sacred, Healing, and Hallucinogenic Powers* (Rochester, N.Y.: Healing Arts Press, 1992), p. 134.
2. J. W. Papez, "A Proposed Mechanism of Emotion," *Archives of Neurology and Psychiatry* (1937) 38:725–743.
3. Schultes and Hofmann, *Plants of the Gods*, p. 176.
4. *Rāja yoga* (Calcutta: Advaita Ashrama, 1994), p. 88.

Chapter 7

1. S. Radhakrishnan, *Search for Truth* (Delhi: Hind Pocket Books, 1995), p. 34.
2. *The Creation of the Universe,* PBS Home Video, North Star Productions, 1985.
3. P. Davies, *The Mind of God* (New York: Touchstone, Simon and Schuster, 1992), p. 176.
4. *The Creation of the Universe,* 1985.
5. Ibid.
6. Quoted by H. V. Dehejia and K. Vatsyayan, *The Advaita of Art* (Delhi: Motilal Banarsidass Publishers, 1996), p. 1.
7. Quoted by N. Ray, *Idea and Image in Indian Art* (Delhi: Munshiram Manoharlal Publishers, 1973), p. 5.
8. C. Darwin, *The Expression of the Emotions in Man and Animals,* 3d ed. (Oxford: Oxford University Press, 1998), pp. 92, 94–95.

9. C. Darwin, *On the Origin of Species by Means of Natural Selection, or the Preservation of Favored Races in the Struggle for Life* (London: Penguin Books, 1985), p. 115.

10. A. C. Doyle, *The Complete Sherlock Holmes* (New York: Doubleday and Company, 1988).

Chapter 8

1. K. H. Potter, R. E. Buswell Jr., P. S. Jaini, and N. R. Reat, *Encyclopedia of Indian Philosophies,* vol. 7, *Abidharma Buddhism to 150 A.D.* (Delhi: Motilal Banarsidass Publishers, 1996), pp. 13–14.

2. M. K. Gandhi, *An Autobiography: The Story of My Experiments with Truth* (Boston: Beacon Press Books, 1957), p. 158.

3. M. Gandhi, *Letters to a Disciple* (New York: Harper and Bros., 1950), p. 224.

4. Ibid., p. 504.

5. C. King, *The Words of Martin Luther King Jr.* (New York: Newmarket Press, 1983).

Chapter 9

1. M. Kaku, *Hyperspace: A Scientific Odyssey Through Parallel Universes, Time Warps, and the Tenth Dimension* (New York: Anchor Books, Doubleday, 1994), p. 9.

2. R. Moody Jr., *Life After Life* (New York: Bantam Books, 1975), p. 58.

3. M. Grey, *Return from Death* (London: Arkana, Penguin, 1985).

Chapter 10

1. *Common Sense About Yoga* (Calcutta: Swapna Printing Works, 1990).

Glossary

Ācārya: Revered teacher
Adharma: Unfair, unethical, and immoral actions
Ādi: Original, first
Advaita Vedānta: Philosophy of nonduality expanded and popularized by Śaṃkara.
Ānanda maya kośa: The part of the mind that generates bliss.
Ānanda: Spiritual bliss. Unlike the mundane sensual pleasure that contrasts with pain, ānanda had no antipode.
Annamaya kośa: The part of the mind that represents the body.
Arthaśāstra: Economics in ancient India.
Aryabhāṭa: The sixth century Indian mathematician.
Āsana: Posture usually related to meditation and yogic exercises.
Āśrama: Hermitage, stage of life.
Āstika: Believer in traditional (Vedic) views of philosophy and religion.
Āśvamedha: Horse sacrifice ritual.
Atharva Veda: The fourth Veda that deals with spells, magic, curses and cures.
Ātmabodha: Self awareness, enlightenment.
Ātman: Soul, divine spark, transcendental consciousness.
Avidya: The mindset that accepts the phenomenal world as "real."

Bāṣya: Commentary
Bhagavad Gītā: Divine songs. One of the most profound and sacred Hindu scriptures.
Bhakti yoga: Union with the Absolute through devotion.
Bhartṛhari. A Sixth Century Indian grammarian and philosopher.
Bodhi: Enlightenment. To be awakened.
Brahma muhūrta: Auspicious time. Usually the early morning hours before the sunrise.
Brahma: The God of creation in the Hindu triumvirate.
Brahman: The Spirit of the Universe, the Absolute, the Divine. Derived from the root Sanskrit word "brh" means to grow or burst forth.

Brahmasūtra: "Aphorisms of about the nature of the Divine" written by Bādarāyaṇa.
Brahmasūtrabāsya: Commentray on Brahmasutra by Śāṃkara.
Brāhmaṇa: The priestly class that occupies the uppermost echelon of the caste hierarchy. The same word denotes parts of the vedas that deal with liturgy.

Caraka Samhita: Textbook of medicine by Caraka written during the early part of the Common Era.
Cārvāka: A school of materialism in ancient India.
Cit: Consciousness that is the basis for all mental activities but is not directly related to any.
Citta: Consciousness specific to different mental acticities.

Dhammapada: The Buddhist canonical text that enjoys universal acceptance as the Buddha's own words.
Dharmaśāstra: Code of ethics and jurisprudence. The best known one was written by Manu around 400 B.C.
Dravidian: The darker skinned people who live in the southern parts of India. Some scholars believe that their ancestors founded the pre-Aryan Indus valley civilization.
Dvāpara yuga: The third of the four ages.

Gạngạ̄: River Ganges.
Gauḍāpada: Sixth century Indian philosopher.
Gāyatrī: Famous mantra from Ṛg Veda.
Gītā: Short for Bhagavad Gita.
Guṇa: Quality, characteristic, attribute. According to the ancient Sāmkhya theorists, matter (Prakṛti) forms into three strains called guṇas.

Hanūmān: Monkey God from Rāmāyana.
Hatha yoga: A system of philosophy that places primary emphasis on physical exercises and postures.

Jaina: A contemporary of the Buddha. His given name was Mahāvīra. He founded Jainism.
Jīva: Soul, life.
Jñāna: Superior knowledge that is intuitive and not mediated by perception or intellect.
Jñāna Kāṇḍa: The school of Hindu theistic philosophy that places primary emphasis on the intellect in gaining spirtual insights and experiences.
Jñāna Yoga: Enlightenment through intuitive and intellectual knowledge.
Jyotiṣāṁ Jyotiḥ: Light of lights.
Jyotiśāstra : Astrology of ancient India.

Kāli: The malefic aspect of the goddess.
Kāli Yoga: The last of the four ages. It is the dark age.
Kānda: Section, part, chapter.
Kārikā: Verse, commentary, treatise.

Karma kānda: The Hindu theistic philosophy that emphasizes rites and rituals.

Karma yoga: Union with the Absolute through conduct.

Kashmir Śaivism: An esoteric system of philosophy of monism with a theistic (Śīva) focus.

Kośa: Sheath, cover.

Kriyā: Action, skill, movement, function.

Kṛṣṇa: Popular Hindu God from Mahābhārata.

Kuṇḍalinī: A system of meditation closely related to Tantrism. The system is based of the abstract concept of a snake that remains coiled within the vertebral column with its tail at the coccyx and its head at the vertex. With progress in meditation, different "centers" on the body of the snake become "alive." Enlightenment occurs when the head comes alive and the normally dichotomous principles of male and female unite.

Lokāyata: A school of materialism in ancient India.

Mādhyāmika: Middle doctrine.

Mahā: Great, revered.

Mahābhārata: The larger of the two Indian Epics.

Mahāvākyā: The most revered, profound and meaningful passages from the Upaniṣads.

Maithuna: Sexual practices considered vulgar, incestuous, adulterous, etc.

Mamsa: Flesh, meat.

Manah: Mind.

Manāna: To think.

Maṇḍalas: Geometrical forms with symbolic significance that enable the aspirant to transcend the material world.

Manomaya kośa. The part of the mind that represents memory, thoughts, etc.

Matsya: Fish.

Māyā: The insubstantial, phantasmal nature of the mundane world.

Māyuraaṇḍa Vāda: The peacock-egg hypothesis. All attributes manifested in the bird have to be present in a latent form in the egg.

Mokṣa: Salvation, release, spiritual freedom.

Mūdrā: Parched grain, fermented cereal. Hand gesture in dance forms.

Nāda: Sound.

Nādabrahman: A term coined by Bhartṛhari who divinized speech. It stand for speech.

Nādyaśāstra: Drama in ancient India.

Nāgārjuna: Buddhist philosopher who developed the highly intellectual Madhyāmaka school of Māhāyana Buddhism.

Nāma: Name.

Nāstika. Atheist, unbeliever, one who does not recognize the Vedas.

Nirvāṇa: Buddhist concept of enlightenment.

Nishkāma karma: Action unmotivated by personal gain.

Niyama: Observance, discipline. From the eight-fold system of philosophy developed by Patañjali.

Nṛttaśāstra: Dance in ancient India.
Nyāya: Logic, axiom, reasoning.

Om: Sacred sound that stands for "everything." Om is etymologically related to such terms as omnipresent, omniscient, and omnipotent.

Padmasāna: Lotus posture.
Pañca-śīla: The five precepts of Buddhism: nonviolence, honesty, sexual continence, truthfulness, and temperance.
Pānini: The fourth century B.C. Sanskrit grammarian.
Patañjali: The second century ascetic who developed the yoga philisophy.
Prajñā: Wisdom, intuitive wisdom, consciousness.
Prakṛti: One (material) component of mind-matter dichotomy in the Sāmkhya philosophy; Prakṛti represents the matter.
Prāṇa: Vital air, life, soul.
Prāṇamaya kośa: The part of the mind that represents the vital functions.
Prāṇāyāma: Breath control. Breathing exercises from the eight-fold system of philosophy developed by Patañjali.
Pratyāhāra : Obliteration of senses. Withdrawal from the manifested world of senses. From the eight-fold system of philosophy developed by Patañjali.
Purāṇa: Hindu mythology.
Puruṣa: One (mind) component of mind-matter dichotomy in the Sāmkhya philosophy; Puruṣa represents the mind.

Rāga: Basic tunes in Indian music; greed, attachment.
Rāja yoga: The royal path to enlightenment.
Rajas: According to the ancient Samkhya theorists, matter (Prakṛti) is constituted by three strains of which rajas represents energy and activity.
Rāma: Hero of the epic Rāmāyaṇa, believed to an incarnation of Viṣṇu.
Rāmāyaṇa: The smaller but earlier and more elegant of the two epics.
Rasa: The intensely blissful, esthetic emotion associated with art.
Rāsaśāstra: Chemistry in ancient India.
Ṛg Veda: The most ancient of Hindu scripture.
Ṛṣi: Ascetic, sage.

Śabda: Sound, word, verbal testimony.
Śabdabrahman: A term coined by Bhartṛhari who divinized speech. The term stands for undifferentiated sound prior to speech.
Śakti: The female principle in Tantrism.
Sālagrāma: Large pebbles from the River Gāndaki, sacred to Viṣṇu.
Sāma Veda: The third of the four Vedas that contains melodious rendition of Ṛg Veda verses.
Samādhi: The enlighted state of mind associated with deep meditation. It is the Hindu equivalent of Buddhist nirvāṇa.

Saṃgham: Buddhist term meaning group or commune.
Śāṁkara: Eighth century Indian philosopher considered to be the main architect of the Advaita (Non dual) school.
Saṁsāra: The manifested world of senses characterized by birth and death, sensual pleasure and pain, and desire and disappointments.
Sannyāsa: Asceticism.
Sannyāsin: Ascetic.
Sanskṛtam: Sanskrit.
Saraswatī : Ṛg Vedic goddess and the patron of all art, especially music.
Sarnath: The place near the present day Banares where the Buddha delivered his first sermon.
Śāstra: Science.
Sat: According to the ancient Sāmkhya theorists, matter (Prakṛti) is constituted by three strains of which sat represents equilibrium between the positive rajas and negative tamas.
Satya agraha: A term Gandhi coined. Satya stands for truth and agraha for firmness. Together they mean firm establishment in truth.
Satyam: Truth.
Satya yoga: The first of the four ages.
Śīrśāsana: A yogic exercise that involves standing on the head for long periods of time.
Śīva: The God of destruction in the Hindu triumvirate.
Śīvam: Grace, elegance.
Smṛti: Sacred texts composed by sages. This is differentiated from the unabridged "word of God." The Buddhist term smṛti has been translated as mindfulness.
Soma: The scared drug of the Vedas that facilitated communion with the Divine, identified recently as Amanita muscaria.
Spanda: Pulsation, vibration. This is a term from Kashmir Śaivism.
Spoḍa: A term coined by Bhartṛhari who divinized speech. Spoda is the mechanism by which speech takes form from inchoate elements.
Sri-chakra: An elegant geometrical design (yantra) composed of overlapping circles, with philosophical, religious, and spiritual significance for many Hindus.
Śruti: Religious texts considered revelatory. "The word of God." Derived from śravana—to hear.
Sthūla- śarīra: Gross physical body.
Sukṣma-śarīra: Subtle body or mind.
Sundaram: Beautiful.
Śūnyatā: Nothingness, emptiness.
Śūnyavāda: The Buddhist doctrine of nothingness.
Śuśruta Samhita: Textbook of surgery by Śuśruta written during the first centuries of the Common Era.
Tamas: According to the ancient Sāmkhya theorists, matter (Prakṛti) is constituted by three strains of which tamas represents inertia and inactivity.
Tanmayībhāva: Total immersion of mind and body in esthetic enjoyment.

Tantrism: The ancient antinomian element in Hinduism that violates all traditional bans. It revolves around union of the male (Śīva) and female (Śakti) elements; the female part is given special emphasis.

Tapas: Tapa literally means "inner heat," and tapas, the austere practices of sages designed to fan the inner "flames." In their forest abodes the sages practice a highly disciplined and simple lifestyle, and their entire lives are spent in such practices as meditation, yoga, and prayer.

Thāṇḍavam: A dance form symbolizing the fury of cataclysmic destruction, associated with Śīva.

Treta yuga: The second of the four ages.

Theravādins: The conservative faction of Buddhists.

Turīya: The fourth stage beyond sleep, dreaming, and wakefulness.

Upaniṣads: Revered religious texts, usually found at the end of the Vedas. They are regarded as intellectually and philosophically the most sophisticated of Hindu scripture.

Vāc: Word.

Vajrāyaṇa: Tantric Buddhism that violates some traditional Buddhist prohibitions.

Vardhamāna: A contemporary of the Buddha who expanded and popularized Jainism.

Varṇa: Color, Caste system.

Veda: Divine knowledge. Most ancient cannonical texts in Hinduism.

Vedanā: Sense experience, sensation, feeling.

Vedānta: The end of Vedas, Upaniṣads.

Vijñāna: Mundane knowledge mediated by the senses or the intellect.

Vijñānamaya kośa: The part of the mind that represents intelligence.

Vishṇu: The God of preservation and protection in the Hindu triumvirate.

Vivekananda: Twentieth century well-known and highly respected Indian philosopher.

Vyākaraṇam: Grammer.

Yagur Veda: The second Veda that deals with liturgy.

Yama: Abstention, restraint, self-abnegation. From the eight-fold system of philosophy developed by Patañjali.

Yantras: Geometrical designs with symbolic significance that enable the aspirant to transcend the material world.

Yoga sūtra: The school of philosophy founded by the second century sage, Patañjali. The eight-fold system requires a highly disciplined and organized lifestyle in addition to the practice of meditation.

Yoga: Derived from the word "yuj"—to unite—yoga means union or confluence. More specifically, it means the union of man with God.

Yuga: Age or cycle, aeon. Indians recognize four yugas.

References

Introduction

Alcoholics Anonymous World Services, Inc. (1976). *Alcoholics Anonymous.* 3d ed. New York.

Bhattacharya, K. (1978). *The Dialectical Method of Nagarjuna. Vigrahavyavartani,* ed. E. H. Johnston and A. Kunst. Delhi: Motilal Banarsidass.

Davids, T.W.R., and H. Oldenberg. (1998). "Vinaya Texts, Part III." In *Sacred Books of the East,* ed. Max F. Müller. Delhi: Motilal Banarsidass Publishers.

Institute of Medicine. (1989). *Prevention and Treatment of Alcohol Problems: Research Opportunities.* Washington, D.C.: National Academy Press

______. (1990). *Broadening the Base of Treatment for Alcohol Problems.* Washington, D.C.: National Academy Press.

James, W. (1991). *The Varieties of Religious Experience.* New York: Triumph Books.

Kessler, R. C., K. A. McGonagle, S. Zhao, C. B. Nelson, M. Hughes, S. Eshleman, H. U. Wittchen, and K. S. Kendler. (1994). "Lifetime and Twelve-Month Prevalence of DSM-III-R Psychiatric Disorders in the United States. Results from the National Comorbidity Survey." *Archives of General Psychiatry* 51:8–19.

Miller, W. R., and B. S. McCrady, eds. (1993). "*The Importance of Research on Alcoholics Anonymous.*" In Research on Alcoholics Anonymous. New Brunswick, N.J.: Rutgers Center of Alcohol Studies.

Nace, E. P. (1987). *The Treatment of Alcoholism.* New York: Brunner-Mazel.

Radhakrishnan, S. (1994). *The Principal Upaniṣads.* New Delhi: Indus, Harper-Collins.

Seventh Special Report to the U.S. Congress on Alcohol and Health. (1990). Rockville, Md.: U.S. Department of Health and Human Services, Public Health Service, Alcohol, Drug Abuse and Mental Health Administration.

Singh, J. (1968). *An Introduction to Madhyamaka Philosophy.* Delhi: Motilal Banarsidass.

Trice, H. M, and W. J. Staudenmeier Jr. (1989). *A Social Cultural History of Alcoholics Anonymous in* Alcoholism. Vol. 7: *Treatment Research,* ed. M. Galanter. New York: Plenum Press.

Vivekananda. (1995). *Bhakti Yoga.* Calcutta: Advaita Ashrama.

Warner, L. A., R. C. Kessler, M. Hughes, J. C. Anthony, and C. B. Nelson. (1995). "Prevalence and Correlates of Drug Use and Dependence in the United States." *Archives of General Psychiatry* 52:219–229.

Chapter 1

Allchin, B., and R. Allchin. (1982). *The Rise of Civilization in India and Pakistan.* Cambridge: Cambridge University Press.

Apte, V. M. "The Vedangas." In *The Cultural Heritage of India.* Vol. 1: *The Early Phases,* ed. S. K. Chatterji, N. Dutt, A. D. Pusalker, N. K. Bose. Calcutta: The Ramakrishna Mission.

Boslough, B. (1989). *Stephen Hawking's Universe and Introduction to the Most Remarkable Scientist of Our Time.* New York: Avon Books.

Chatterji, S. K. (1958). "Linguistic Survey of India: Languages and Scripts." In *The Cultural Heritage of India.* Vol. 1: *The Early Phases,* ed. S. K. Chatterji, N. Dutt, A. D. Pusalker, N. K. Bose. Calcutta: The Ramakrishna Mission.

Davies, P. (1992). *The Mind of God: The Scientific Basis for a Rational World.* New York: Simon and Schuster.

Doniger, W. (1991). *The Laws of Manu.* London: Harmondsworth and Penguin.

Dutta, B. (1986). "Vedic Mathematics." In *The Cultural Heritage of India.* Vol. 6: *Science and Technology,* ed. P. Ray and S. N. Sen. Calcutta: The Ramakrishna Mission.

Gavin, F. (1996). *An Introduction to Hinduism.* Cambridge: Cambridge University Press.

Khuswant, Singh. (1974). *India: An Introduction.* New Delhi: Vision Books.

Leaky, R., and R. Lewin. (1995). *The Sixth Extinction: Pattern of Life and the Future of Humankind.* New York: Doubleday.

Michio, K., and J. Thompson. (1995). *Beyond Einstein: The Cosmic Quest for the Theory of the Universe.* New York: Anchor Books.

Mookerji, R. (1962). *Asoka.* Delhi: Motilal Banarsidass Publishers.

Nehru, J. (1981). *The Discovery of India.* New Delhi: Jawaharlal Nehru Memorial Fund, dist. by Oxford University Press.

Parpola, A. (1994). *Deciphering the Indus Script.* Cambridge: Cambridge University Press.

Penrose, R. (1989). *The Emperor's New Mind: Concerning Computers, Minds, and Laws of Physics.* Oxford: Oxford University Press.

Radhakrishnan, S. (1989). *Indian Philosophy.* Vol. 1. Delhi: Oxford University Press.

______. (1993). *The Bhagavad Gita.* New Delhi: Indus, HarperCollins.

———. (1994). *The Principal Upaniṣads*. New Delhi: Indus, HarperCollins.
Sagan, C. (1980). *Cosmos*. New York: Random House.
Sen, S. N., and A. K. Bag. (1986). "Post-Vedic Mathematics." In *The Cultural Heritage of India*. Vol. 6: *Science and Technology*, ed. P. Ray and S. N. Sen. Calcutta: The Ramakrishna Mission.
Sen Gupta, P. C. (1986). "Astronomy in Ancient India." In *The Cultural Heritage of India*. Vol. 6: *Science and Technology*, ed. P. Ray and S. N. Sen. Calcutta: The Ramakrishna Mission.

Chapter 2

Benson, D. F. (1994). *The Neurology of Thinking*. New York: Oxford University Press.
Critchley, E., ed. (1995). *The Neurological Boundaries of Reality*. Northvale, N.J.: Jason Aronson.
Davids, R.T.W. (2000). *Dialogues of the Buddha*, 1st Indian ed., Sāmaṇṇphala Sutta. Delhi: Motilal Banarsidass.
Davids, T.W.R., and H. Oldenberg. (1998). "Vinaya Texts, Part II." In *Sacred Books of the East*, ed. Max F. Müller. Delhi: Motilal Banarsidass Publishers.
Fischbach, G. D. (1992). "Mind and Brain." *Scientific American* (September):48–76.
Fish, F. (1967). *Clinical Psychopathology: Signs and Symptoms in Psychiatry*. Bristol: John Wright and Sons.
Granville-Grossman, K. (1971). *Recent Advances in Clinical Psychiatry*. Edinburgh and London: Churchill.
Kaku, M., and J. Thompson. (1995). *Beyond Einstein*. New York: Anchor Books.
Purves, D., G. J. Augustine, D. Fitzpatrick, and L. C. Katz. (1997). *Neuroscience*, ed. A. S. Lamantia and J. O. McNamara. Sunderland, Mass.: Sinanuer Associates Publishers.
Shepherd, G. M. (1983). *Neurobiology*. New York: Oxford University Press.
Singh, J. (1968). *An Introduction to Madhyamaka Philosophy*. Delhi: Motilal Banarsidass.

Chapter 3

Asayeva, N. (1993). *Shankara and Indian Philosophy*. Albany: State University of New York Press.
______. (1995). *From Early Vedanta to Kashmir Shaivism*. Albany: State University of New York Press.
Einstein, A. (1961). *Relativity: The Special and the General Theory*, authorized trans. Robert W. Lawson. New York: Three Rivers Press.
Fish, Frank. (1967). *Clinical Psychopathology: Signs and Symptoms in Psychiatry*. Bristol: John Wright and Sons.

Frydman, M. (1976). *I Am That: Talks with Sri Nisargadatta Maharaj,* 2d ed. Durham, N.C.: Acron Press.

Hawking, S. (1988). *A Brief History of Time: From the Big Bang to Black Holes.* New York: Bantam Books.

______. (1993). *Black Holes and Baby Universes and Other Essays.* New York: Bantam Books.

Kaku, M., and J. Thompson. (1987). *Beyond Einstein: The Cosmic Quest for the Theory of the Universe.* New York: Anchor Books.

O'Flaherty, W. D. (1991). *The Rig Veda: An Anthology.* London: Penguin Books.

Radhakrishnan, S. (1923). *Indian Philosophy.* Vol. 1. Calcutta: Oxford University Press.

______. (1994). *The Principal Upaniṣads.* New Delhi: Indus, HarperCollins.

Shree Purohit Swami, and Yeats, W. B. (1992). *The Ten Principal Upaniṣhads.* Calcutta: Rupa and Co.

Swāmi Gambhīrānanda. (1957). *Eight Upaniṣads.* Vol. 1 (translation). Calcutta: Advaita Ashrama.

______. (1986). *Śvetāśvatara Upaniṣad* (translation). Calcutta: Advaita Ashrama.

______. (1989). *Eight Upaniṣads.* Vol. 2 (translation). Calcutta: Advaita Ashrama.

Chapter 4

Appleton, J. P. (1993). "The Contribution of the Amygdala to Normal and Abnormal Emotional States." *Trends in Neuroscience* 16:328–333.

Benson, D. F. (1994). *The Neurology of Thinking.* Oxford: Oxford University Press.

Cotman, C. W., and J. L. McGaugh. (1980). *Behavioral Neuroscience: An Introduction.* New York: Academic Press.

Damasio, A. R., and H. Damasio. (1992). "Brain and Language." *Scientific American* 2 (3) (September):67.

Daniel, D. G., R. J. Mathew, and W. H. Wilson. (1988). "Sex Roles and Regional Cerebral Blood Flow." *Psychiatry Research* 27:55–64.

Darwin, C. (1998). *The Expression of the Emotions in Man and Animals.* 3d ed. Oxford: Oxford University Press.

Dasgupta, S. N. (1975). *A History of Indian Philosophy.* Vol. 6 (Indian edition). New Delhi: Cambridge University Press and Motilal Banarsidass.

Davis, M. (1992). "The Role of the Amygdala in Fear and Anxiety." *Annual Review of Neuroscience* 15:353–375.

De Santis, S. (1995). *Nature and Man: The Hindu Perspectives.* Vol. 1. : Varanasi, Sociecos, and Dilip Kumar Publishers.

Descartes, R. Torrey, H.A.P. (1892). *The Philosophy of Descartes in Extracts from His Writings.* New York: Holt.

Deutsch, G., and H. M. Eisenberg. (1987). "Frontal Blood Flow Changes in Recovery from Coma." *Journal of Cerebral Blood Flow and Metabolism* 7:29–34.

Deutsche, van Buitenen, J. A. B. (1971). *A Source Book of Advaita Vedanta.* Honolulu: University Press of Hawaii.

Easwaran, E. (1985). *The Bhagavad Gita.* Tomales, Calif.: Nilgiri Press.

Feirtag, M., and W.J.H. Nauta. (1986). *Fundamental Neuroanatomy.* New York: W. H. Freeman.

Fuster, J. M. (1980). *The Prefrontal Cortex. Anatomy, Physiology, and Neuropsychology of the Frontal Lobe.* New York: Raven Press.

Gazzaniga, M. S. (1998). *The Mind's Past.* Berkeley: University of California Press.

Goldman-Rakic, P. S. (1984). The Frontal Lobes: Uncharted Provinces of the Brain. *Trends in Neuroscience* 7:425–429.

Gupta, B. (1995). *Perceiving in Advaita Vedanta: Epistemological Analysis and Interpretation.* New Delhi: Motilal Banarsidass.

Ingvar, D. H. (1979). "Hyperfrontal Distribution of the Cerebral Gray Matter Flow in Resting Wakefulness: On the Functional Anatomy of the Conscious State." *Acta Neurol Scand* 60:12–25.

Ingvar, D. H., and U. Soderberg. (1958). "Cortical Blood Flow Related to EEG Patterns Evoked by Stimulation of the Brain Stem." *Acta Physiol Scand* 42:130–143.

Juge, O., J. S. Meyer, M. Sakai, F. Yamaguchi, M. Yamamoto, and T. Shaw. (1979). "Critical Appraisal of Cerebral Blood Flow Measured from Brain Stem and Cerebellar Regions After ^{133}Xe Inhalation in Humans." *Stroke* 10:428–437.

Kimura, D. (1992). "Sex Differences in the Brain." *Scientific American* 2 (3) (September):67.

Kluver, H., and P. C. Bucy. (1939). "Preliminary Analysis of Functions of the Temporal Lobes in Monkeys." *Archives of Neurology and Psychiatry* 42:979–1000.

Larson, G. J., and R. S. Bhattacharya. (1987). "Samkhya." In *Encyclopedia of Indian Philosophies,* vol. 5. New Delhi: Motilal Banarsidass.

Lishman, W. A. (1987). *Organic Psychiatry. The Psychological Consequences of Cerebral Disorder.* 2d ed. Oxford: Blackwell Scientific Publications.

Loewy, A. D., and K. M. Spyer. (1990). *Central Regulation of Autonomic Functions.* Oxford: Oxford University Press.

Luria, A. (1973). *The Working Brain: An Introduction to Neuropsychology.* New York: Basic Books.

Mathew, R. J. (1989). "Hyperfrontality of Regional Cerebral Blood Flow Distribution in Normals During Resting Wakefulness: Fact or Artifact?" *Biological Psychiatry* 26:717–724.

Mathew, R. J., W. H. Wilson, and D. G. Daniel. (1985). "The Effect of Non-Sedating Doses of Diazepam on Regional Cerebral Blood Flow." *Biological Psychiatry* 20:1109–1116.

Mathew, R. J., W. H. Wilson, and S. R. Tant. (1986). "Determinants of Resting Regional Cerebral Blood Flow in Normal Subjects." *Biological Psychiatry* 21:907–914.

Miller, G. A. (1965). "Some Preliminaries to Psycholinguistics." *American Psychologist* 20:15–20.

Moruzzi, G., and H. W. Magoun. (1949). "Brain Stem Reticular Formation and Activation of the EEG." *Electroencephalography and Clinical Neurophysiology* 1:455–473.

Myers, D. G., and E. Diener. (1997). "The Pursuit of Happiness: Mysteries of the Mind." *Scientific American Special Issue.*

Pande, G. C. (1994). *Life and Thought of Śaṅkarācārya.* New Delhi: Motilal Banarsidass Publishers.

Papez, J. W. (1937). "A Proposed Mechanism of Emotion." *Archives of Neurology and Psychiatry* 38:725–743.

Penfield, W., and G. Mathieson. (1974). "Memory." *Archives of Neurology* 31:145–147.

Potter, K. H., R. E. Buswell Jr., P. S. Jaini, and N. R. Reat. (1996). "Abhidharma Buddhism." In *Encyclopedia of Indian Philosophies,* vol. 7. New Delhi: Motilal Banarsidass Publishers.

Purves, D., G. J. Augustine, D. Fitzpatrick, L. C. Katz, A. S. LaMantia, and J. O. McNamara. (1997). *Neuroscience.* Sunderland, Mass.: Sinauer Associates.

Radhakrishnan, S. (1923). *Indian Philosophy.* Vol. 2. Calcutta: Oxford University Press.

______. (1994). *The Principal Upaniṣads.* New Delhi: Indus, HarperCollins.

Restak, R. M. (1984). *The Brain.* New York: Bantam Books.

Seal, B. (1895). *The Positive Sciences of the Ancient Hindus.* New Delhi: Motilal Banarsidass Publishers.

Shepherd, G. M. (1983). *Neurobiology.* Oxford: Oxford University Press.

Svāmī Turīyānanda. (1987). *Vivekacūdāmaṇī of Srī Saṅkarācārya* (translation). Madras: Sri Ramakrishna Math.

Swāmi Gambhīrānanda. (1957). *Eight Upaniṣads.* Vol. 1 (translation). Calcutta: Advaita Ashrama.

Chapter 5

Bates, D., and N. Cartlidge. (1994). "Disorders of Consciousness." In *The Neurological Boundaries of Reality,* ed. E.M.R. Critchley. London: Jason Aronson.

Bhattacharya, K. (1978). *The Dialectical Method of Nagarjuna: Vigrahavyavartani,* ed. E. H. Johnston and A. Kunst. Delhi: Motilal Banarsidass.

Buchsbaum, M. S., J. C. Gillin, J. Wu, et al. (1989). "Regional Cerebral Glucose Metabolic Rate in Human Sleep Assessed by Positron Emission Tomography." *Life Science* 45:1349–1346.

Chalmers, D. J. (1996). *The Conscious Mind: In Search of a Fundamental Theory.* Oxford: Oxford University Press.

Crick, F. (1994). *The Astonishing Hypothesis: The Scientific Search for the Soul.* New York: Simon and Schuster.

Crick, F., and C. Koch. (1992). "The Problem of Consciousness." *Scientific American* (September):153–159.

Davidson, J. M., and R. J. Davidson. (1980). *The Psychobiology of Consciousness.* New York: Plenum Press.

Dewhurst, K. (1982). *Hughlings Jackson on Psychiatry.* Oxford: Sandford Publications.

Hobson, J. A. (1989). *Sleep.* New York: Scientific American Library.

Indich, W. M. (1980). *Consciousness in Advaita Vedanta.* New Delhi: Motilal Banarsidass.

Kennedy, C., J. C. Gillin, W. Mendelson, et al. (1982). "Local Cerebral Glucose Utilization in Non-Rapid Eye Movement Sleep." *Nature* 297:325–327.

King, J., and K. H. Pribram, eds. (1995). *Scale in Unconscious Experience: Is the Brain Too Important to Be Left to Specialists to Study.* Mahwah, N.J.: Lawrence Erlbaum Associates.

Madsen, P. L., and S. Vorstrup. (1991). "Cerebral Blood Flow and Metabolism During Sleep." *Cerebrovascular and Brain Metabolism Reviews* 3:281–296.

Obrist, W. D., T. A. Jennarelli, H. Segawa, C. A. Dolinskas, and T. W. Lingfitt. (1979). "Relation of Cerebral Blood Flow to Neurological Status and Outcome in Head-Injured Patients." *Journal of Neurosurgery* 51:292–300.

Parthasarathy, A. (1971). *Atmabodha: By Sri Adi Sankaracharya,* (translation). Bombay: Vedanta Life Institute.

Potter, K., R. E. Buswell Jr., P. S. Jaini, and N. R. Reat. (1996). "Abhidharma Buddhism to 150 A.D." In *Encyclopedia of Indian Philosophies,* vol. 7. New Delhi: Motilal Banarsidass.

Radhakrishnan, S. (1994). *The Principal Upaniṣads.* New Delhi: Indus, HarperCollins.

Singh, J. (1968). *An Introduction to Madhyamaka Philosophy.* Delhi: Motilal Banarsidass.

Svāmī Turkyānanda. (1987). *Vivekacūdāmaṇī of Srī Saṅkarācārya* (translation). Madras: Sri Ramakrishna Math.

Swami Gambhirananda. (1993). *Brahma-Sūtra-Bhāsya of Srī Sankarācārya* (translation). Calcutta: Advaita Ashrama.

Swāmi Jagadānanda. (1941). *Upadeśa Sāhasrī of Srī Sankarāchārya* (translation). Madras: Sri Ramakrishna Math.

Chapter 6

Abel, E. L. (1980). *Marijuana: The First 12,000 Years.* New York: Plenum Press.

American Psychiatric Association. (1994). *Diagnostic and Statistical Manual of Mental Disorders.* 4th ed. Washington, D.C.: American Psychiatric Association.

Artigues, M., and R. J. Mathew. (1996). "Alcohol Intoxication and Homicide." *Medicine and Law* 15:485–491.

Banks, A., and T.A.N. Waller. (1988). *Drug Misuse: A Practical Handbook for GPs.* Oxford: Blackwell Scientific Publications.

Bates, D., and N. Cartlidge. (1995). *Disorders of Consciousness.* In *The Neurological Boundaries of Reality,* ed. E.M.R. Critchley. London: Jason Aronson.

Bell, R., H. Wechsler., and L. D. Johnston. (1997). "Correlates of College Student Marijuana Use: Results of a U.S. National Survey Addiction." 92:571–581.

Bergman, R. L. (1971). "Navajo Peyote Use: Its Apparent Safety." *American Journal of Psychiatry* 128:51–55.

Brands, B., B. Sproule, and J. Marshman. (1998). *Drugs and Drug Abuse.* 3d ed. Ontario: Addiction Research Foundation.

Brawley, P., and J. C. Duffield. (1972). "The Pharmacology of Hallucinogens." *Pharmacological Reviews* 24:31–66.

Brust, J.C.M. (1993). *Neurological Aspects of Substance Abuse.* Boston: Butterworth-Heinemann.

Clark, L. D., R. Hughes, and E. N. Nakashima. (1970). "Behavioral Effects of Marijuana: Experimental Studies." *Archives of General Psychiatry* 23:193–198.

Cooper, J. R., F. E. Bloom, and R. H. Roth. (1982). *The Biochemical Basis of Neuropharmacology.* 4th ed. Oxford: Oxford University Press.

Davids, T.W.R., and H. Oldenberg. (1998). "Vinaya Texts, Part II." In *Sacred Books of the East,* ed. Max F. Müller. Delhi: Motilal Banarsidass Publishers.

______. (1998). "Vinaya Texts, Part III." In *Sacred Books of the East,* ed. Max F. Müller. Delhi: Motilal Banarsidass Publishers.

Dixon, J. C. (1963). "Depersonalization Phenomena in a Sample Population of College Students." *British Journal Of Psychiatry* 109:371–375.

Feirtag, M., and W. J. H. Nauta. (1986). *Fundamental Neuroanatomy.* New York: W. H. Freeman and Co.

Flood, G. (1996). *An Introduction to Hinduism.* Cambridge: Cambridge University Press.

Fuster, J. M. (1980). *The Prefrontal Cortex: Anatomy, Physiology, and Neuropsychology of the Frontal Lobe.* New York: Raven Press.

Hartmann, D. E. (1988). *Neuropsychological Toxicology: Identification and Assessment of Human Neurotoxic Syndromes.* New York: Pergamon Press.

Harwood-Nuss, A., guest ed. (1990). *Emergency Aspects of Alcoholism: Emergency Medicine Clinics of North America.* Vol. 8, no. 4. Philadelphia: W. B. Saunders Company.

Hermle L., M. Fuenfgeld, G. Oepen, H. Botsch, D. Borchardt, E. Gouzoulis, R. A. Fehrenbach, and M. Spitzer. (1992). "Mescaline-Induced Psychopathological, Neuropsychological and Neurometabolic Effects in Normal Subjects: Experimental Psychosis as a Tool for Psychiatric Research." *Biological Psychiatry* 32:976–991.

Hollister, L. E. (1986). "Health Aspects of Cannabis." *Pharmacological Reviews* 38:1–20.

______. (1988). "Cocaine—1988." *Human Psychopharmacology* 3:1–2.

Huxley, A. (1963). *The Doors of Perception.* New York: Harper and Row Publishers.

Ivry, R. B., and S. W. Keele. (1989). "Timing Functions of the Cerebellum." *Journal of Cognitive Neuroscience* 1:136–152.

______. (1993). "Cerebellar Involvement in the Explicit Representation of Temporal Information." *Annals of New York Academy of Sciences* 682:214–230.

Ivry, R. B., S. W. Keele, and H. C. Diener. (1988). "Dissociation of the Lateral and Medium Cerebellum in Movement Timing and Movement Execution." *Experimental Brain Research* 73:167–180.

Johnston, E. H. (1984). *Aśvaghoṣa's Buddhacarita or Acts of the Buddha*. Delhi: Motilal Banarsidass Publishers.

Jolly, J. (1992). "The Institutes of Vishṇu." In *Sacred Books of the East*, ed. Max F. Müller. Delhi: Motilal Banarsidass Publishers.

Keele, S. W., and R. B. Ivry. (1991). "Does the Cerebellum Provide a Common Computation for Diverse Tasks? A Timing Hypothesis." *Annals of New York Academy of Sciences* 608:179–211.

Kelly, D. (1976). "Neurosurgical Treatment of Psychiatric Disorders." In *Recent Advances in Clinical Psychiatry*, ed. K. Granville-Grossman. London: Churchill Livingstone.

Kenna, J. C., and G. Sedman. (1964). "The Subjective Experience of Time During Lysergic Acid Diethylamide (LSD–25) Intoxication." *Psychopharmacologia* 5:280–288.

Kinsley, D. R. (1986). *Hindu Goddesses*. Berkeley: University of California Press.

Lader, M. H. (1969). "Psychophysiological Aspects of Anxiety." In *Studies of Anxiety*, ed. M. H. Lader. Ashford, Kent: Hedley Brothers (English Language Rights Royal Medico-Psychological Association).

Lishman, W. A. (1987). *Organic Psychiatry: The Psychological Consequences of Cerebral Disorder*. 2d ed. Oxford: Blackwell Scientific Publications.

Logan, W. J. (1975). "Neurological Aspects of Hallucinogenic Drugs." In *Advances in Neurology*, vol. 13, ed. W. J. Friedlander. New York: Raven Press.

Luria, A. R. (1973). *The Working Brain: An Introduction to Neuropsychology*. New York: Basic Books.

Mathew, R. J., W. H. Wilson, R. E. Coleman, T. G. Turkington, and T. R. DeGrado. (1997). "Marijuana Intoxication and Brain Activation in Marijuana Smokers." *Life Sciences* 60:2075–2089.

Mathew, R. J., W. H. Wilson, D. F. Humphreys, J. V. Lowe, and K. E. Weithe. (1992). "Regional Cerebral Blood Flow After Marijuana Smoking." *Journal of Cerebral Blood Flow and Metabolism* 12:750–758.

______. (1993). "Depersonalization After Marijuana Smoking." *Biological Psychiatry* 33:431–441.

Mathew, R. J., W. H. Wilson, T. G. Turkington, and R. E. Coleman. (1998). "Cerebellar Activity and Disturbed Time Sense After THC." *Brain Research* 797:183–189.

Mayer-Gross, W. (1935). "On Depersonalization." *British Journal of Medical Psychology* 15:8–122.

McKim, W. A. (1991). *Drugs and Behavior: An Introduction to Behavioral Pharmacology.* New Jersey: Prentice-Hall.

Mechoulam, R. (1986). Introduction. In *Cannabinoids as Therapeutic Agents,* ed. R. Mechoulam. Boca Raton, Fla.: CRC Press.

Melges, F. T., J. R. Tinklenberg, M. Deardorff, N. H. Davis, R. E. Anderson, and C. A. Owen. (1974). "Temporal Disorganization and Delusional-Like Ideation." *Archives of General Psychiatry* 30:855–861.

Melges, F. T., J. R. Tinklenberg, L. E. Hollister, and H. K. Gillespie. (1970a). "Marijuana and Temporal Disintegration." *Science* 168:118–120.

______. (1970b). "Temporal Disintegration and Depersonalization During Marijuana Intoxication." *Archives of General Psychiatry* 23:204–210.

Melges, F. T., J. R. Tinklenberg, L. E. Hollister, and H. K. Gillespie. (1971). "Marijuana and the Temporal Span of Awareness." *Archives of General Psychiatry* 24:564–567.

Mellor, C. S. (1988). "Depersonalization and Self-Perception." *British Journal of Psychiatry* 153 (supplement 2):15–19).

Myers, D. H., and G. Grant. (1972). "A Study of Depersonalization in Students." *British Journal of Psychiatry* 12:59–65.

Nahas, G. G., D. J. Harvey, and M. Paris. (1984). *Marijuana in Science and Medicine.* New York: Raven Press.

Nauta, W.J.H. (1971). "The Problem of Frontal Lobe: A Re-Interpretation." *Journal of Psychiatric Research* 8:167–187.

Noyes, R., Jr., and R. Kletti. (1977). "Depersonalization Response to Life-Threatening Danger." *Comprehensive Psychiatry* 18:375–384.

O'Flaherty, W. D. (1981). *The Rig Veda* (translation). London: Penguin Books.

Oepen, G., M. Fuenfgeld, A. Harrington, L. Hermle, and H. Botsch. (1989). "Right Hemisphere Involvement in Mescaline-Induced Psychosis." *Psychiatry Research* 29:335–336.

Patel, A. R. (1968). "Mescaline and Related Compounds." *Progress in Drug Research,* pp. 11–47.

Papez, J. W. (1937). "A Proposed Mechanism of Emotion." *Archives of Neurology and Psychiatry* 38:725–743.

Radhakrishnan, S. (1994). *The Principal Upaniṣads.* New Delhi: Indus, HarperCollins.

Redda, K. K., C. A. Walker, and G. Barnett. (1989). *Cocaine, Marijuana, Designer Drugs: Chemistry, Pharmacology, and Behavior.* Boca Raton, Fla.: CRC Press.

Reeves, H., J. DeRosnay, Y. Coppens, and D. Simonnet. (1996). *Origins: Cosmos, Earth, and Mankind.* New York: Arcade Publishing.

Riedlinger, T. J. (1993). "Wasson's Alternative Candidates for Soma." *Journal of Psychoactive Drugs* 25:149–156.

Robert, W. W. (1960). "Normal and Abnormal Depersonalization." *Journal of Mental Science* 106:478–493.

Roth, M., and N. Argyle. (1988). "Anxiety, Panic, and Phobic Disorders: An Overview." *Journal of Psychiatric Research* 22:33–54.

Schultes, R. E., and A. Hofmann. (1992). *Plants of the Gods: Their Sacred, Healing, and Hallucinogenic Powers.* Rochester, N.Y.: Healing Arts Press.

Scott, M. (1983). *Kundalini in the Physical World.* London: Arkana, Penguin.

Sedman, G. (1966). "Depersonalization in a Group of Normal Subjects." *British Journal of Psychiatry* 112:907–912.

______. (1970). "Theories of Depersonalization: A Re-Appraisal." *British Journal of Psychiatry* 117:1–14.

Shulgin, A. T. (1973). "Mescaline: The Chemistry and Pharmacology of Its Analogs." *Lloydia* 36:45–58.

Stafford, P. (1978). *Psychedelics Encyclopedia.* 3d expanded ed. Berkeley: Ronin Publishing.

Strassman, R. J. (1984). "Adverse Reactions to Psychedelic Drugs: A Review of the Literature." *Journal of Nervous and Mental Disease* 172:577–595.

______. (1994). "Human Psychopharmacology of LSD, Dimethyl Tryptamine and Related Compounds in 50 years of LSD: Current Status and Perspectives of Hallucinogens." In *A Symposium of the Swiss Academy of Medical Sciences,* ed. A. Pletscher and D. Ladewig. New York: Parthenon Publishing Company.

______. (1995). "Hallucinogenic Drugs in Psychiatric Research and Treatment." *Journal of Nervous and Mental Disease* 183:127–138.

Svoboda, R. E. (1992). *Ayurveda, Life, Health, and Longevity.* London: Arkana, Penguin.

Tinklenberg, J. R., W. T. Roth, and B. S. Kopell. (1976). "Marijuana and Ethanol: Differential Effects of Time Perception, Heart Rate, and Subject Response." *Psychopharmacology* 49:275–279.

Washton, A. M. and M. S. Gold. (1987). *Cocaine: A Clinician's Handbook.* New York: Guilford Press.

Chapter 7

Ahern, G. L. (1995). "Cerebral Laterality and Consciousness (In Reply)." *Archives of Neurology* 51:337–338.

Albert, M. L., R. Silvelberg, A. Recher, and M. Berman. (1976). "Cerebral Dominance for Consciousness." *Archives of Neurology* 33:453–454.

Asayeva, N. (1995). *From Early Vedanta to Kashmir Shaivism.* Albany: State University of New York Press.

Benowitz, L. I., D. M. Bear, R. Rosenthal, M. M. Mesulam, E. Zaidel, and R. W. Sperry. (1983). "Hemispheric Specialization in Nonverbal Communication." *Cortex* 19:5–11.

Bever, T. G. (1988). "A Cognitive Theory of Emotion and Aesthetics in Music." *Psychomusicology* 7:165–175.

Bogen, J. E., and H. W. Gordon. (1974). "Musical Tests for Functional Lateralization with Intracarotid Amobarbital." *Nature* 230:524.

Bose, S. (1990). *Indian Classical Music.* New Delhi: Vikas Paperbacks.

Confavreux, C., B. Croisile, P. Garassus, G. Aimand, and M. Trillet. (1992). "Progressive Amusia and Aprosody." *Archives of Neurology* 49:971–976.
Cunningham, J. C. (1988). "Developmental Change in the Understanding of Affective Meaning in Music." *Motivation and Emotion* 12:399–413.
Darwin, C. (1998). *The Expression of the Emotions in Man and Animals.* 3d ed. Oxford: Oxford University Press.
Davids, T.W.R. (1994). "Buddhist Suttas." In *Sacred Books of the East,* ed. Max F. Müller. Delhi: Motilal Banarsidass Publishers.
Davies, P. (1992). *The Mind of God: The Scientific Basis for a Rational World.* New York: Simon and Schuster.
Dehejia, H. V. (1996). *The Advaita of Art.* New Delhi: Motilal Banarsidass.
Dennis, M. (1980). "Capacity and Strategy for Syntactic Comprehension After Right or Left Hemidecortication." *Brain and Language* 10:287–317.
Doyle, A. C. (1988). *The Complete Sherlock Holmes.* New York: Doubleday and Company.
Easwaran, E. (1985). *The Bhagavad Gita.* Tomales, Calif.: Nilgiri Press.
Eccles, J. C. (1989). *Evolution of the Brain: Creation of the Self.* London: Routledge.
Fenwick, P. (1995). "Alterations in Conscious Awareness." In *The Neurological Boundaries of Reality,* ed. E.M.R. Critchley. Northvale, N.J.: Jason Aronson.
Flood, G. (1996). *An Introduction to Hinduism.* Cambridge: Cambridge University Press.
Hellige, J. B. (1993). *Hemispheric Asymmetry. What's Right and What's Left?* Cambridge: Harvard University Press.
Ivry, R. B., and L. C. Robertson. (1998). *The Two Sides of Perception.* Cambridge: MIT Press.
Joseph, R. (1992). *The Right Brain and the Unconscious: Discovering the Stranger Within.* New York: Plenum Press.
Kelsey, M. (1981). *Tongue Speaking: The History and Meaning of Charismatic Experience.* New York: Cross Road.
Kinsley, D. R. (1988). *Hindu Goddesses.* Berkeley: University of California Press.
Mithen, S. (1996). *The Prehistory of the Mind: The Cognitive Origins of Art, Religion and Science.* London: Thames and Hudson.
Myers, P. S. (1999). *Right Hemisphere Damage: Disorders of Communication and Cognition.* San Diego: Singular Publishing Group.
O'Flaherty, W. D. (1981). *The Rig Veda* (translation). London: Penguin Classics.
Ottoson, D., ed. (1987). *Duality and Unity of the Brain.* London: Macmillan Press.
Pande, G. C. (1994). *The Life and Thought of Saṅkārācārya.* New Delhi: Motilal Banarsidass.
Pinker, S. (1997). *How the Mind Works.* New York: W. W. Norton and Company.
Potter, K. H. (1981). "Advaita Vedānta up to Śaṃkara and His Pupils." In vol. 3 of *Encyclopedia of Indian Philosophies.* New Delhi: Motilal Banarsidass Publishers.
______. (1996). "Abhidharma Buddhism to 150 A.D." In vol. 7 of *Encyclopedia of Indian Philosophies.* New Delhi: Motilal Banarsidass.

Radhakrishnan, S. (1923). *Indian Philosophy.* Vol. 1. New Delhi: Oxford University Press.

_______. (1994). *The Principal Upaniṣads.* New Delhi: Indus, HarperCollins.

Ramachandran, V. S., and S. Blakeslee. (1998). *Phantoms in the Brain.* New York: William Morrow and Company.

Rao, S.K.R. (1989). *Sri Chakra.* New Delhi: Sri Satguru Publications.

Ray, N. (1973). *Idea and Image in Indian Art.* New Delhi: Munshiram Manoharlal Publishers.

Reeves, H., J. D. Rosnay, Y. Coppens, and D. Simonnet. (1996). *Origins, Cosmos, Earth, and Mankind.* New York: Arcade Publishing.

Ridley, M., ed. (1997). *Evolution.* Oxford: Oxford University Press.

Rosadine, G., and G. F. Rossi. (1967). "On the Suggested Cerebral Dominance for Consciousness." *Brain* 90:101–112.

Serafetinides, E. A. (1995). "Cerebral Laterality and Consciousness." *Archives of Neurology* 52:337.

Serafetinides, E. A., R. D. Hoare, and M. V. Driver. (1965). "Intracarotid Sodium Amylobarbitone and Cerebral Dominance for Speech and Consciousness." *Brain* 88:107–130.

Shepherd, G. M. (1983). *Neurobiology.* Oxford: Oxford University Press.

Smith, A. (1966). "Speech and Other Functions After Left (Dominant) Hemispherectomy." *Journal of Neurology, Neurosurgery, and Psychiatry* 29:467–471.

Swami Gambhirananda. (1957). *Eight Upaniṣads.* Vol. 1 (translation). Calcutta: Advaita Ashrama.

Tagore, R. (1931). *The Religion of Man.* London: Unwin Books.

Thompson, W. F., and B. Robitaille. (1992). "Can Composers Express Emotions Through Music?" *Empirical Studies of the Arts* 10:79–89.

Tucker, D. M., (1987). "Hemispheric Specialization: A Mechanism for Unifying Anterior and Posterior Brain Regions." In *Duality and Unity of the Brain,* ed. D. Ottoson. London: Macmillan Press.

Tucker, D. M., and P. A. Williamson. (1984). "Asymmetric Neural Control Systems in Human Self-Regulation." *Psychological Review* 91:185–215.

Warrier, A.G.K. (1993). *Srīmad Bhagavad Gītā Bhāṣya of Sri Saṁkarācārya.* Madras: Sri Ramakrishna Math.

Zatorre, R. J., and A. R. Halpern. (1993). "Effect of Unilateral Temporal-Lobe Excision on Perception and Imagery of Songs." *Neuropsychologia* 31:221–232.

Chapter 8

Alabaster, H. (1996). *The Wheel of the Law: Buddhism.* New Delhi: Motilal Banarsidass Publishers.

Bjorklunde, Lindvall A. (1978). *The Meso-Telencephalic Dopamine Neuron System: A Review of Its Anatomy in Limbic Mechanisms,* ed. K. Livingston and Hornykiewiczo. New York: Plenum Press.

Coward, H. G. and K. K. Raja. (1990). "The Philosophy of the Grammarians." In *Encyclopedia of Indian Philosophies*, vol. 5. New Delhi: Motilal Banarsidass.

Doniger, W., and B. K. Smith. (1991). *The Laws of Manu*. London: Penguin Books.

Easwaran, E. (1985). *The Bhagavad Gita*. Tomales, Calif.: Nilgiri Press.

______. (1996). *The Dhammapada*. Tomales, Calif.: Nilgiri Press.

Gandhi, M. K. (1957). *An Autobiography: The Story of My Experiments with Truth*. Boston: Beacon Press Books.

Gazzaniga, M. S. (1998). *The Mind's Past*. Berkeley: University of California Press.

Griffith, R.T.H. (1995). *Hymns of the Rigveda*. New Delhi: Motilal Banarsidass Publishers.

Hellige, J. B. (1993). *Hemispheric Asymmetry: What's Right and What's Left?* Cambridge: Harvard University Press.

Iyer, R. (1990). *The Essential Writings of Mahatma Gandhi*. New Delhi: Oxford University Press.

Koob, G. F. (1992). "Drugs of Abuse: Anatomy, Pharmacology and Function of Reward Pathways." *Ti PS* 13 (May):117–184.

Lammens, H.S.J. (1998). *Islam: Beliefs and Institutions*. New Delhi: Motilal Banarsidass Publishers.

Maqsood, D. R. (1994). *Islam*. Lincolnwood, Ill.: NTC Publishing Group.

Mathews, J. K. (1989). *The Matchless Weapon Satyagraha*. Bombay: Bharatiya Vidya Bhavan.

Merton, T. (1964). *Gandhi on Nonviolence*. New York: New Directions Publishing Corporation.

Muscaro, J. (1973). *The Dhammapada, the Path of Perfection*. London: Penguin Books.

Potter, K. H. (1996). "Abhidharma Buddhism to 150 A.D." *Encyclopedia of Indian Philosophies*, vol. 7. Delhi: Motilal Banarsidass Publishers.

Radhakrishnan, S. (1923). *Indian Philosophy*. Vol. 1. Calcutta: Oxford University Press.

______. (1994). *The Principal Upaniṣads*. New Delhi: Indus, HarperCollins.

Simpson, W. (1996). *The Buddhist Praying-Wheel*. New Delhi: Motilal Banarsidass.

Skinner, B. F. (1935). "Two Types of Conditioned Reflex and a Pseudotype." *Journal of General Psychology* 12:66–77.

Stoddart, W., and R. A. Nicholson. (1998). *Sufism—the Mystical Doctrines and the Idea of Personality*. New Delhi: Motilal Banarsidass.

Thomas, E. J. (1996). *Early Buddhist Scriptures*. New Delhi: Motilal Banarsidass.

Tucker, D. M. (1987). "Hemispheric Specialization: A Mechanism for Unifying Anterior and Posterior Brain Regions." In *Duality and Unity of the Brain*, ed. D. Ottoson. New York: Plenum Press.

Tucker, D. M., and P. A. Williamson. (1984). "Asymmetric Neural Control Systems in Human Self-Regulation." *Psychological Review* 91:185–215.

Wise, R. A. (1981). "Intracranial Self-Stimulation: Mapping Against the Lateral Boundaries of the Dopaminergic Cells of the Substantia Nigra." *Brain Research* 213:190–194.

______. (1990). "The Role of Reward Pathways in the Development of Drug Dependence: Psychotropic Drugs of Abuse." In *International Encyclopedia of Pharmacology and Therapeutics*, ed. J. K. Balfourd, section 130. New York: Pergamon Press.

Wise, R. A., and M. A. Bozarth. (1984). "Brain Reward Circuitry: Four Crucial Elements 'Wired' in Apparent Series." *Brain Research Bulletin* 12:203–208.

Chapter 9

Anand, B. K., G. S. Chhina, and B. Singh. (1961). "Some Aspects of EEG Studies in Yogis." *EEG and Clinical Neurophysiology* 13:452–456.

Arberry, A. J. (1972). *Sufism: An Account of the Mystics of Islam.* London: Allen and Unwin.

Banquet, J. P. (1973). "Spectral Analysis of the EEG in Meditation." *Electroencephalography and Clinical Neurophysiology* 5:143–151.

Benson, H. (1975). *The Relaxation Response.* New York: William Morrow.

Bhikku, S. (1949). *The Way of Mindfulness.* Colombo, Ceylon: Vajirama.

Bloomfield, M. (1972). *The Kausika Sutra of Atharva Veda.* New Delhi: Motilal Banarsidass.

______. (1987–1992). *Hymns of the Atharva Veda.* New Delhi: Motilal Banarsidass.

Buhler, G. (1988–1993). *The Laws of Manu* (translation). New Delhi: Motilal Banarsidass.

Chhina, G. S., and B. Singh. (1975). "The State of Research in Yoga." *Science Today* (June). Bombay: Times of India Press.

Conze, E. (1956). *Buddhist Meditation.* London: Allen and Unwin.

Dalai Lama XIV (1965). *An Introduction to Buddhism.* New Delhi: Tibet House.

Dasgupta, S. N. (1989). *A Study of Patanjali.* New Delhi: Motilal Banarsidass.

Dass, R. (1978). *Journey of Awakening.* New York: Bantam Books.

Davids, R.T.W. (2000). *Dialogues of the Buddha,* 1st Indian ed., Sāmayyphala Sutta. Delhi: Motilal Banarsidass.

Davidson, J. M. (1976). "The Physiology of Meditation and Mystical States of Consciousness." *Perspectives in Biology and Medicine* 19:345–379.

Eggeling, J. (1972–1976). *The Satapatha-Brahmana.* New Delhi: Motilal Banarsidass.

Fort, A. O. (1990). *The Self and Its States.* New Delhi: Motilal Banarsidass.

Frost, G., and Y. Frost. (1989). *Tantric Yoga: The Royal Path to Raising Kundalini Power.* York Beach, Maine: Samuel Weiser.

Ganapati, S. V. (1992). *Sama Veda.* New Delhi: Motilal Banarsidass.

Goleman, D. (1988). *The Meditative Mind.* New York: G. P. Putnam's Sons.

Grey, M. (1985). *Return from Death.* London: Arkana, Penguin.

Griffith, R.T.H. (1995). *Hymns of the Rig Veda.* New Delhi: Motilal Banarsidass.

Halevi, Z.B.S. (1976). *The Way of Kabbalah.* New York: Samuel Weiser.

Hazra, R. C. (1987). *Studies in the Puranic Records on Hindu Rights and Customs.* New Delhi: Motilal Banarsidass.

Kasamatsu, A., and T. Hirai. (1966). "An Electroencephalographic Study of the Zen Meditation (Zazen)." *Folio Psychiatrica Neurologica Japonica* 20:315–316.

______. (1969). *An EEG Study on the Zen Meditation in Altered States of Consciousness*, ed. C. Tart. New York: Wiley.

Kornfield, J. (1993). *A Path with Heart: A Guide Through the Perils and Promises of Spiritual Life.* New York: Bantam Books.

Kubler-Ross, E. (1970). *On Death and Dying.* London: Tavistock Publications.

Lal, B. B. (1998). *New Light on the Indus Valley Civilization.* New Delhi: Aryan Books International, Motilal Banarsidass Publishers.

Matsuoka, S. (1990). "Theta Rhythms: State of Consciousness." *Brain Topography* 3:203–208.

Merton, T. (1960). *The Wisdom of the Desert.* New York: New Directions.

Moody, R. A., Jr. (1975). *Life After Life.* New York: Bantam Books.

Muller, F. M. (1981–1995). *The Upaniṣads.* New Delhi: Motilal Banarsidass.

Peers, A. (1958). *St. John of the Cross Ascent of Mt. Carmel* (translation). Garden City, N.Y.: Image Books.

Potter, K. H., ed. (1996). "Abhidharma Buddhism to 150 A.D." In *Encylopedia of Indian Philosophies,* vol. 7. New Delhi: Motilal Banarsidass.

Radhakrishnan, S. (1994). *The Principal Upaniṣads.* New Delhi: Indus, Harper-Collins.

Scholem, G. (1974). *Kabbalah, Quandrangle.* New York: Times Book Company.

Shastri, J. L. (1978–1995). *The Sivapurana.* New Delhi: Motilal Banarsidass.

______. (1996). *Manu Smrti.* New Delhi: Motilal Banarsidass.

Stoddart, W., and R. A. Nicholson. (1998). *Sufism—the Mystical Doctrines and the Idea of Personality.* New Delhi: Aryan Books International, Motilal Banarsidass Publishers.

Sāṁkhya-yogāchārya Swāmi Hariharānanda Āraṇya (P. N. Mukerji). (1983). *Yoga Philosophy of Patañjali* (translation). Albany: State University of New York Press.

Swami Prabhavananda, and C. Isherwood. (1994). *Patanjali Yoga Sutras* (translation). Madras: Sri Ramakrishna Math.

Venkatesh, S., T. R. Raju, V. Y. Shivani, G. Tompkins, and B. L. Meti. (1997). "A Study of Structure of Phenomenology of Consciousness in Meditative and Non-Meditative States." *India Journal of Physiology and Pharmacology* 41:149–153.

Wallace, R. K. (1970). "Physiological Effects of Transcendental Meditation." *Science* 167:1751–1754.

Wallace, R. K., H. Benson, and A. Wilson. (1971). "A Wakeful Hypometabolic Physiologic State." *American Journal of Physiology* 221:795–799.

West, M. (1979). "Meditation." *British Journal of Psychiatry* 135:457–467.

Chapter 10

Carrazana, E., J. DeToledo, W. Tatum, R. Rivas-Vasquez, G. Rey, and S. Wheeler. (1999). "Epilepsy and Religious Experiences: Voodoo Possession." *Epilepsia* 40:239–241.

Cirignotta, F., C. V. Todesco, and E. Lugarasi. (1980). "Temporal Lobe Epilepsy with Ecstatic Seizures (So-Called Dostoyevsky Epilepsy)." *Epilepsia* 21:705–710.

Csernansky, J. G., D. B. Liederman, M. Mandabach, and J. A. Moses Jr. (1990). "Psychopathology and Limbic Epilepsy: Relationship to Seizure Variables and Neuropsychological Function." *Epilepsia* 31:275–180.

Dewhurst, K., and W. Beard. (1970). "Sudden Religious Conversion in Temporal Lobe Epilepsy." *British Journal of Psychiatry* 117:497–509.

Foote-Smith, E., and L. Bayne. (1991). "Joan of Arc." *Epilepsia* 32:810–815.

Foote-Smith, E., and T. J. Smith. (1996). "Emanuel Swedenborg." *Epilepsia* 37:211–218.

Hay, D. (1990). *Religious Experience Today: Studying the Facts.* London: Mowbray.

Landsborough, D. (1987). "St. Paul and Temporal Lobe Epilepsy." *Journal of Neuorology, Neurosurgery, and Psychiatry* 50:659–664.

Ogata, A., and T. Miyakawa. (1998). "Religious Experiences in Epileptic Patients with a Focus on Ictus-Related Episodes." *Psychiatry and Clinical Neurosciences* 52:321–325.

Index